Introduction to the Technology of Explosives

Paul W. Cooper
Stanley R. Kurowski

New York • Chichester • Weinheim • Brisbane • Singapore • Toronto

The procedures in this text are intended for use only by persons with prior training in the field of explosives. In the checking and editing of these procedures, every effort has been made to identify potentially hazardous steps and safety precautions have been inserted where appropriate. However, these procedures must be conducted at one's own risk. The authors and the publisher, its subsidiaries and distributors, assume no liability and make no guarantees or warranties, express or implied, for the accuracy of the contents of this book or the use of information, methods or products described in this book. In no event shall the authors, the publisher, its subsidiaries or distributors, be liable for any damages or expenses, including consequential damages and expenses, resulting from the use of the information, methods or products described in this book.

Paul W. Cooper
424 Girard Blvd., SE
Albuquerque, NM 87106

Stanley R. Kurowski
3212 La Veta Dr., NE
Albuquerque, NM 87110

This book is printed on acid-free paper. ∞

Library of Congress Cataloging-in-Publication Data
Cooper, Paul W., 1937–
Introduction to the technology of explosives / Paul W. Cooper, Stanley R. Kurowski.
p. cm.
Includes bibliographical references and index.
ISBN 0-471-18635-X (alk. paper)
1. Explosives. I. Kurowski, Stanley R. II. Title.
TP270.C744 1966
662'.2—dc20 95-52613
CIP

To order books or for customer service please, call 1(800)-CALL-WILEY (225-5945).

Printed in the United States of America

ISBN 0-471-18635-X Wiley-VCH, Inc.

20 19 18

To Judi and Carol,
thank you.

Preface

About This Book

Introduction to the Technology of Explosives deals with the various technologies used in the design and application of explosives and explosive systems. The intent of the book is twofold: (1) to provide the inexperienced worker in the field with sufficient background to understand problems that may arise and to facilitate interaction with specialists in the field; and (2) to provide an awareness of the crucial importance of safety in dealing with explosives to be obtained (one hopes) through a combination of technical knowledge and an appreciation of the various laws, statutes, codes, and common practices in this field. The book also provides the basic calculational skills needed to solve simple, first-order engineering design problems (the majority of problems in engineering applications can be solved with first-order, or "back-of-the-envelope" type calculations).

This text briefly covers the basic areas of the chemistry of explosives, thermochemistry, shock and detonation, initiation theory, and criteria for detonators. In addition, the book covers the use of scaling in the analysis and design of explosive systems, explosive devices, and instrumentation used in the explosive laboratories. Numerous graphs, tables, illustrations, and equations are included to help the reader understand and use explosive engineering technology. The text concludes with an overview of the legal and industrial codes and requirements related to the manufacture, storage, transportation, and use of explosive materials, devices, and systems.

The text is arranged so that the subject matter follows a logical progressive sequence, so that the information being presented will be more understandable for readers new to this field. In Chapter 1 the reader is introduced to basic explosive chemistry and pertinent chemical reactions. Explosive materials are categorized by chemical type as well as by how the explosives are used. The same is done for propellants and pyrotechnics. Chapter 2 explains how to control the burning of propellants and how to calculate the state of the product gases in order to determine the input parameters required for interior ballistics. Chapter 3 takes the reader from the basics of sound propagation through nonreactive shock wave behavior to simple detonation theory. Chapter 4 introduces the reader to the theories of initiation and ignition sensitivity of explosives and shows how these apply to the design and behavior of nonelectric, hot-wire, exploding-wire, and slapper detonators. Chapter 5 introduces the concept of scale modeling and how scaling is used in designing explosive systems and as an analytical tool in predicting explosive effects such as air and water blast and crater formation. Chapter 6 defines and illustrates many of the explosive components and devices that are available "off-the-shelf." Chapter 7 gives the reader some insight into the various codes and requirements for the transportation and storage of explosives. Chapter 8 leads the reader through various requirements and recommendations for explosives facilities and operations.

Some Definitions

The term *explosives* generally is used in reference to a wide range of energetic materials that can react chemically to produce heat, light, and gas. This is a very broad definition and, for our purposes, explosives must be further classified. Explosives may be separated into two major categories: *high explosives* and *low explosives*, categories that still are rather broad. High explosives are chemicals that can detonate. High explosives may be further classified into *primary* (those that can be made to detonate very easily), *secondary* (those that are more difficult to detonate), and *tertiary* (those that are the most difficult to detonate) explosives. Tertiary explosives are also referred to as *insensitive high explosives*, or IHEs, which include some blasting agents. Secondary explosives differ from primary explosives in three basic ways:

1. Secondaries do not easily go from burning (deflagration) to detonation.
2. Electrostatic ignition is difficult with secondaries.
3. Secondary ignitions require larger shocks.

Low explosives are classified as *pyrotechnics*, *propellants*, or *blasting agents* that cannot be caused to detonate by means of a common blasting cap. Pyrotechnics are materials that, when burned, produce useful heat, light, smoke, delays, gas, and/or sound. The word *pyrotechnic* is rooted in the Greek words meaning the manipulating of fire (*pyro*, fire; *technic*, manipulating of fire or fire technology). Propellants are materials that burn to produce gases used to perform mechanical work, such as propelling a projectile or pushing a piston. Although propellants are always categorized as such, they also are pyrotechnics. The distinction between the two types of low explosives stems from their primary function. Propellants are manufactured in large quantities and mainly are used to produce gas for small- and large-bore guns and rocket propulsion.

The terms *burn* and *detonate* are used thoughout the book. Both these chemical reaction processes involve oxidation. In burning, or deflagrating, the oxidation takes place relatively slowly. The burn front propagates through the burning material at less than the velocity of sound. The term *deflagration* is used in explosive technology when referring to the deflagration-to-detonation transition (DDT). DDT occurs when a high explosive transfers from a burning reaction to a detonation in a column of explosive. Detonation is a reaction similar to deflagration that occurs at a much faster rate. Detonation is characterized by wave propagation at a supersonic rate with respect to the unreacted material. The propagation rates of detonation range from as low as 1 km/s to as high as 9 km/s, depending upon the density and chemical properties of the explosive.

An *explosion* is a large-scale, noisy, rapid expansion of matter into a volume much greater than its original volume. This can be achieved by (1) bursting a vessel containing a pressurized fluid; (2) rapid heating of air and plasma by an electric arc; (3) a very fast burning reaction; or (4) detonating an explosive material.

A *shock wave* is a high-pressure wave or pressure disturbance that moves through a material at a speed faster than the speed of sound in that material. The disturbance is not smooth or continuous. Shock waves can stress a material beyond its elastic limits so that the material will not return to its original state after release of the pressure.

A *sound wave* is a longitudinal pressure wave moving through a material at pressures lower than those of shock waves. The pressures associated with sound waves do not stress the material beyond its elastic limit.

Fragments and *shrapnel* are missiles, e.g. from casing, earth, building materials, etc. When an explosion occurs, fragments may be torn from the explosion and scattered. Shrapnel is a preformed individual piece or pieces that are placed in or around the explosive to produce lethal missiles for military purposes.

Acknowledgments

The authors wish to thank and acknowledge many people who helped with bringing this book to completion. First our thanks to Glenda Ponder for the editing, typing, and formatting of the original manuscript. Our sincere thanks and appreciation to the following people who reviewed the manuscript and provided us with many excellent comments and improvements: Dr. Olden L. Burchett (Sandia National Laboratories, retired), Dr. James E. Kennedy (Los Alamos National Laboratory), Dr. Gerald Laib (Naval Surface Warfare Center White

Oak), Mr. J. Christopher Ronay (Institute of Makers of Explosives), Dr. Ronald Varosh (Reynolds Industries Systems, Inc.), Dr. Brigita M. Dobratz (Lawrence Livermore National Laboratory, retired), Dr. Carl-Otto Lieber (Bundesinstitut fur Chemisch-Technische, BICT, Germany), Dr. Hugh R. James (Atomic Weapons Establishment, England), Dr. Pascal A. Bauer (Professor, Ecole Nationale Superieure de Mecanique et d'Aerotechnique, Paris, France), Dr. Eric J. Rinehart (Field Command, U.S. Defense Nuclear Agency), Mr. Joseph P. Grubisic (deceased, formerly Commander, Bomb and Arson Section, Department of Police, City of Chicago), Lt. Steve Goyen and Detective Lee Prentice (U.S. Army and Los Angeles Police Department, respectively).

Paul W. Cooper
Stanley R. Kurowski
Albuquerque, NM

Contents

CHAPTER

1

Chemistry of Explosives

This chapter will cover the chemical reactions that explosives undergo to produce heat, light, and gases. We will examine the "oxygen balance," a condition that allows us to quantify the degree of reaction. Finally, we will see what explosives look like and how they are catagorized in relation to their chemical makeup and structure. We will also see how these chemical explosives are mixed with other materials to control their physical properties and hence, their mechanical usefulness. In addition to high explosives, we will also examine the chemistry and forms of both propellants and pyrotechnics.

1.1 Chemical Reactions

1.1.1 Exothermic and Endothermic Reactions

A chemical reaction may be considered as the conversion of reactants to products, symbolically written as reactants → products. The internal energy of the reactants is different from the internal energy of the products. Internal energy is energy that is contained in the bonds between the atoms in each molecule. If the reactants contained more energy than the products, then that energy may be released as heat during the reaction. This is called an *exothermic reaction*. The reactions involved in burning and detonation are exothermic. In the opposite case, when the products contain more internal energy than the reactants, that energy has to have been added in order for the reaction to occur. Such a reaction is called an *endothermic reaction*.

1.1.2 Oxidation

There are two major types of reactions involved in burning and detonation. In the first type, there are two reactants, a fuel and an oxidizer, that react to form the products. A simple example of this, shown in Figure 1.1, is the combustion (oxidation) of methane with oxygen. To completely react one methane molecule it requires two oxygen molecules (four oxygen atoms).

Equation (1.1) shows the reactants (one methane molecule, two oxygen molecules, and the products). More generally, the oxidizer does not have to be pure oxygen; it could be an oxidizing salt like a nitrate or perchlorate.

$$CH_4 + 2\ O_2 \rightarrow CO_2 + 2\ H_2O \tag{1.1}$$

In order to burn rapidly, the fuel and oxidizer must be mixed intimately. The smaller the particle sizes (for solids), the faster the burning reaction proceeds. As an example consider the difference between the burning of a wooden log and a dust explosion.

In the second type of reaction, which is by far the more common one in explosives, the oxidizer and fuel are contained within the same starting molecule. This molecule decomposes during the reaction and reforms into oxidized products, as shown in Eq. (1.2). An example of this is the nitroglycol explosive.

$$O_2N{-}O{-}CH_2{-}CH_2{-}O{-}NO_2 \rightarrow 2\ CO_2 + 2\ H_2O + N_2 \tag{1.2}$$

Figures 1.2 and 1.3 show the fuel and oxidizer parts of the nitroglycol molecule, respectively. Note that the fuel portion consists of carbon and hydrogen and the oxidizer portion is composed of the two nitrate esters, a group of explosives that will be discussed later. Notice that this explosive molecule consists of carbon, hydrogen, nitrogen, and oxygen. The great majority of explosives and propellants consist of single molecules made up of these four elements only. These are called CHNO explosives.

Many oxidizer groups and many basic forms of the fuel portion of the explosive molecule will be described later. The molecules depicted above are represented in their structural form. They can also be represented in catalog, or elemental, form. The same nitroglycol molecule can be represented by $C_2H_4N_2O_6$.

The general formula for all CHNO explosives is $C_cH_hN_nO_o$, where c, h, n, and o are the number of carbon, hydrogen, nitrogen, and oxygen atoms, respec-

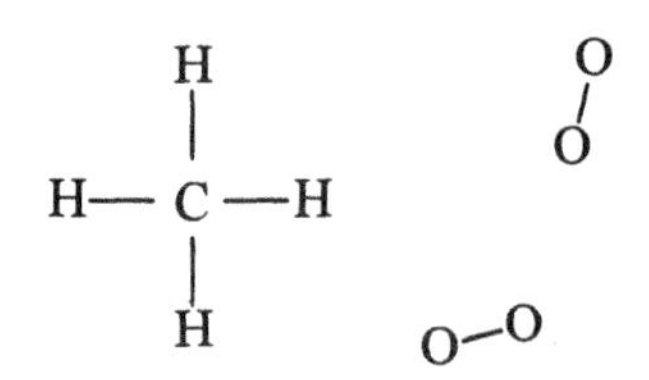

Figure 1.1. The methane molecule and the oxygen required for complete combustion.

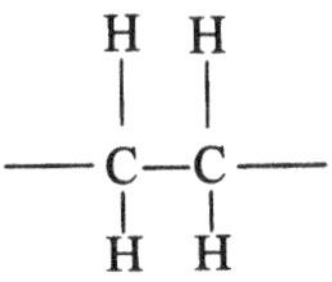

$$
\begin{array}{c}
\text{H} \quad \text{H} \\
| \quad\; | \\
\text{—C—C—} \\
| \quad\; | \\
\text{H} \quad \text{H}
\end{array}
$$

Figure 1.2. Fuel portion of the nitroglycol molecule.

tively, in one molecule of explosive. The simplest picture of how the decomposition reaction occurs is to imagine that in the zone where a propellant is burning or an explosive is detonating, the reactant molecule is completely broken down into its individual component atoms

$$C_cH_hN_nO_o \rightarrow c\,C + h\,H + n\,N + o\,O \tag{1.3}$$

These atoms then recombine to form the ultimate products of the reaction. The typical products formed are

$$
\begin{aligned}
2N &\rightarrow N_2 \\
2H + O &\rightarrow H_2O \\
C + O &\rightarrow CO \\
CO + O &\rightarrow CO_2
\end{aligned} \tag{1.4}
$$

In the case of nitroglycol, as noted above, there is exactly enough oxygen to burn all the carbon completely to carbon dioxide, CO_2. This is not the case with every explosive. Some explosives have more than enough oxygen to burn all the carbon to CO_2. These explosives are called overoxidized or fuel lean. Most explosive compounds do not have enough oxygen to burn all the carbon to CO_2; they are called underoxidized or fuel rich. In all cases, the products formed can be estimated by applying the following "rules of thumb":

1. First all the nitrogen forms N_2.
2. Then all the hydrogen is burned to H_2O.
3. Any oxygen left after H_2O formation burns carbon to CO.
4. Any oxygen left after CO formation burns CO to CO_2.
5. Any oxygen left after CO_2 formation forms O_2.
6. Traces of NO_x (mixed oxides of nitrogen) are always formed.

This set of rules is called the simple product hierarchy for CHNO explosives (and propellants). If the explosive had contained metal additives, these would probably not oxidize until all the above steps were completed. The application

$$O_2N\text{—}O\text{—} \quad \text{and} \quad \text{—}O\text{—}NO_2$$

Figure 1.3. Oxidizer portion of the nitroglycol molecule.

of this simple product hierarchy to three different CHNO explosives is shown in the following examples:

***Example 1.1* Decomposition of nitroglycerine**

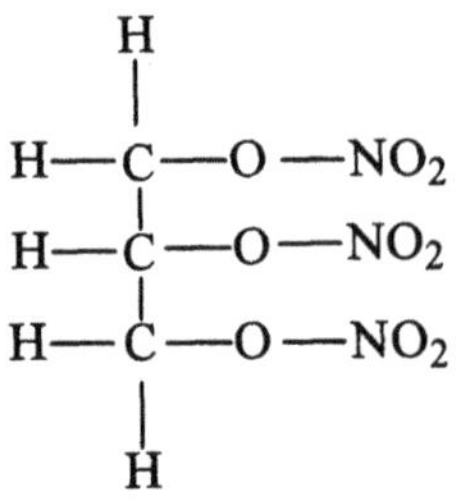

```
     H
     |
H—C—O—NO2
     |
H—C—O—NO2
     |
H—C—O—NO2
     |
     H
```

The oxidizing reaction is as follows:

$$C_3H_5N_3O_9 \rightarrow 3C + 5H + 3N + 9O$$

a. $3N \rightarrow 1.5N_2$
b. $5H + 2.5O \rightarrow 2.5H_2O$ (6.5 O remaining)
c. $3C + 3O \rightarrow 3CO$ (3.5 O remaining)
d. $3CO + 3O \rightarrow 3CO_2$ (0.5 O remaining)

(Note that 8.5 of the 9 atoms of oxygen available have been used. All the H has been burned to H_2O and all the C to CO_2, but there remains one-half an oxygen atom.)

e. $0.5O \rightarrow 0.25O_2$
f. Trace amounts of NO_x

The overall reaction is:

$$C_3H_5N_3O_9 \rightarrow 1.5N_2 + 2.5H_2O + 3CO_2 + 0.25O_2$$

Since there is oxygen remaining, nitroglycerine must be overoxidized.

***Example 1.2* Decomposition of RDX**

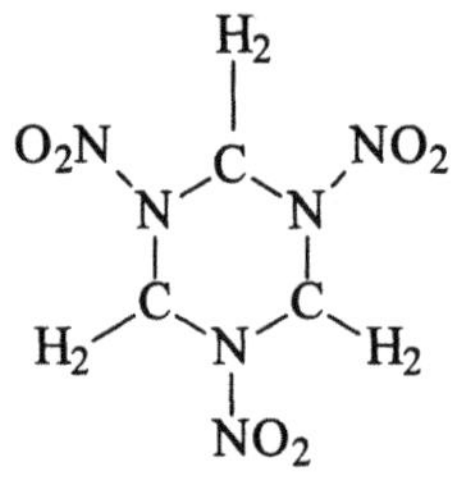

```
             H2
             |
    O2N      C      NO2
       \   /   \   /
         N       N
         |       |
         C       C
    H2 /   \   /   \ H2
             N
             |
            NO2
```

The oxidizing reaction is as follows:

$$C_3H_6N_6O_6 \rightarrow 3C + 6H + 6N + 6O$$

a. $6N \rightarrow 3N_2$
b. $6H + 3O \rightarrow 3H_2O$ (3 O remaining)
c. $3C + 3O \rightarrow 3CO$ (all the O is used at this point; therefore, no CO_2 is formed)

The overall reaction is:

$$C_3H_6N_6O_6 \rightarrow 3H_2O + 3CO + 3N_2.$$

Since there is CO remaining in the products, there was not enough oxygen to completely burn all of the fuel in this molecule; therefore, RDX must be undcroxidized.

Example 1.3 **Decomposition of TNT**

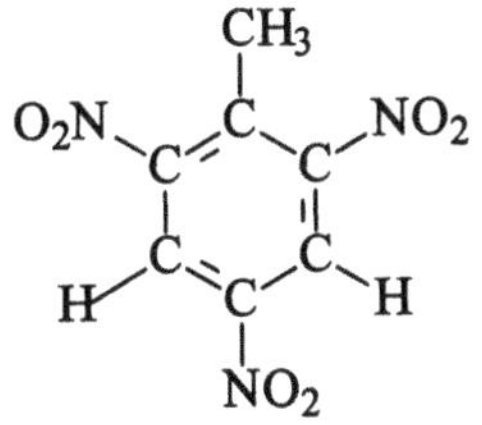

CH_3

O_2N C C C NO_2

H C C C H

NO_2

The oxidizing reaction is as follows:

$$C_7H_5N_3O_6 \rightarrow 7C + 5H + 3N + 6O$$

a. $3N \rightarrow 1.5N_2$
b. $5H + 2.5O \rightarrow 2.5H_2O$ (3.5 O remaining)
c. $7C + 3.5O \rightarrow 3.5CO$ (all the O is used) + 3.5C

The overall reaction is:

$$C_7H_5N_2O_6 \rightarrow 1.5N_2 + 2.5H_2O + 3.5CO + 3.5C$$

Since there is still fuel remaining both C and CO, TNT must be very underoxidized.

In the last two example reactions, some of the products are fuels, specifically, the free carbon, C, and the carbon monoxide, CO. After the burning or detonation reaction is complete, the products may be free to expand into air. As this occurs, they mix with the oxygen in the air, and when the proper mixture with the air is reached, they may burst into flame and burn to CO_2. This second reaction is called a secondary fireball. Such fireballs can also be fueled from other combustible materials that have been combined with the explosive. Casings, glues, binders, and colorants are examples of these burnable materials.

Thus we have concluded that the relative amount of oxygen in an explosive is quite important. When an explosive is exactly balanced, neither rich nor lean, it produces the maximum energy output. The relative amount of oxygen in an explosive (or propellant) is quantitatively expressed as oxygen balance.

1.1.3 Oxygen Balance

Referring to the general formula for a CHNO explosive or propellant, $C_cH_hN_nO_o$, we see that if all the carbon could be burned to CO_2, the number of oxygen atoms required would be twice the number of carbon atoms. Similarly, in order to burn all the hydrogen to H_2O, one oxygen atom would be required for every two hydrogen atoms. For this compound to be exactly balanced, it would need 2c + h/2 atoms of oxygen. It has o atoms of oxygen. Therefore, the quantity

$$o - 2c - h/2 \qquad (1.5)$$

is a measure of the oxygen balance for this molecule. If that number is negative, it means o is less oxygen than is needed for complete combustion; therefore, the material is underoxidized. If o is greater than (2c + h/2), the quantity is positive, which means there is more than enough oxygen, and the molecule is overoxidized.

A similar balance could be derived for burning to CO instead of CO_2, but is not given here. The oxygen balance is a useful parameter in many practical applications. It can be correlated to other properties of burning systems such as the gas volume produced as well as to the heat of explosion.

It is customary to express the oxygen balance (OB) in terms of the weight percent of excess oxygen compared to the weight of explosive. This is done by multiplying the expression (o − 2c − h/2), which is in atom numbers, by the atomic weight (AW) of oxygen, and dividing by the molecular weight (MW) of the explosive material. Then it is put in percent terms by multiplying by a hundred. Thus

$$OB = 100 \frac{AW_o}{MW_{\text{exp}}} [o - (2c + h/2)] \qquad (1.6)$$

The atomic weight of oxygen is 16.000; therefore,

$$OB(\%) = \frac{1600}{MW_{\text{exp}}} [o - (2c + h/2)] \qquad (1.7)$$

The molecular weight of the explosive molecule is the sum of the weights of all the atoms. Since we know the formula is $C_cH_hN_nO_o$, deriving the molecular weight is simple. Table 1.1 gives the atomic weights for the four elements used in CHNO explosives. Therefore, the molecular weight of the explosive is

$$MW_{\text{exp}} = 12.01c + 1.008h + 14.008n + 16o \qquad (1.8)$$

Table 1.1 Atomic Weights for Elements in CHNO Explosives

Chemical Element	Atomic Weight
Carbon	12.010
Hydrogen	1.008
Nitrogen	14.008
Oxygen	16.000

***Example 1.4* Calculating oxygen balance**

Let us calculate the oxygen balance for the four explosives examined earlier.

a. Nitroglycol: $C_2H_4N_2O_6$

$$c = 2, h = 4, n = 2, \text{ and } o = 6$$

$$MW_{exp} = 12.01(2) + 1.008(4) + 14.008(2) + 16.000(6) = 152.068$$

$$OB = \frac{1600}{152.068}\left[6 - 2(2) - \frac{4}{2}\right] = 0\% \qquad \text{(a)}$$

Nitroglycerol is perfectly balanced.

b. Nitroglycerine: $C_3H_5N_3O_9$

$$c = 3, h = 5, n = 3, \text{ and } o = 9$$

$$MW_{exp} = 12.01(3) + 1.008(5) + 14.008(3) + 16.000(9) = 227.094$$

$$OB = \frac{1600}{227.094}\left[9 - 2(3) - \frac{5}{2}\right] = 3.52\% \qquad \text{(b)}$$

Nitroglycerine is slightly overoxidized.

c. RDX: $C_3H_6N_6O_6$

$$c = 3, h = 6, n = 6, \text{ and } o = 6$$

$$MW_{exp} = 12.01(3) + 1.008(6) + 14.008(6) + 16.000(6) = 222.126$$

$$OB = \frac{1600}{222.126}\left[6 - 2(3) - \frac{6}{2}\right] = -21.61\% \qquad \text{(c)}$$

RDX is underoxidized.

d. TNT: $C_7H_5N_3O_6$

$$c = 7, h = 5, n = 3, \text{ and } o = 6$$

$$MW_{exp} = 12.01(7) + 1.008(5) + 14.008(3) + 16.000(6) = 227.134$$

$$OB = \frac{1600}{227.134}\left[6 - 2(7) - \frac{5}{2}\right] = -73.97\% \qquad \text{(d)}$$

TNT is very underoxidized.

To summarize, we have now established that explosive materials are burning chemical systems that contain both the fuel and the oxidizer required for combustion. Some explosive materials contain more fuel than needed, and others are deficient. The degree of excess or deficiency of fuel can be determined by examining the potential products formed in the burning reaction by use of the simple product hierarchy and can be quantified by calculating the oxygen balance. This can be accomplished merely by knowing the chemical formula of the explosive material and the products of the burning reaction.

1.2 Categories of Explosives by Chemical Type

From this point on, we will use the term *explosive* to mean high explosive as defined earlier, that is, those explosives that can detonate. A detonation is a burning or decomposition reaction whose propagation velocity is greater than the speed of sound in the explosive material.

In order to detonate, the explosive must not only react fast but must also produce gas and heat (or energy). Thus the three criteria necessary for a detonation are:

- Reaction propagation velocity faster than bulk sound speed,
- Production of gas, and
- Production of heat.

If any of these three criteria is not met, a detonation cannot occur. We will examine the rates of detonation later; however, the range of detonation velocities for most explosives is between 1 and 9 km/s. Most explosives produce between 500 and 1000 cm^3 of gas for each gram that reacts. Heat produced is normally within the range of 400 to 1200 cal/g reacted. It should be noted that some propellants and pyrotechnic materials will detonate under certain conditions. Later in the text these numbers will be used in detonation calculations.

Many materials can be made to detonate. See Figure 1.4. These materials are composed of either pure or mixed compounds. There are two categories of pure compounds that are determined by their chemical structure. The pure compounds are organic (which are further divided into aromatic and aliphatic structures) and inorganic.

1.2.1 Pure Compounds: Organic

Almost all the explosive materials with which we deal are organic compounds. Organic means that the molecule is built on a skeleton of carbon atoms. These

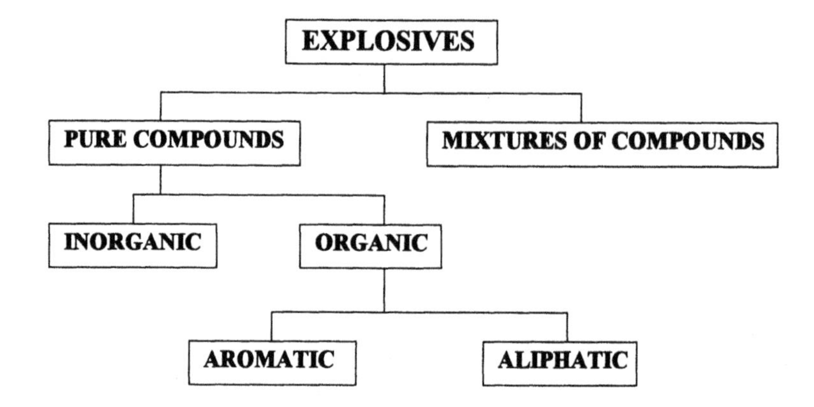

Figure 1.4. Types of explosives.

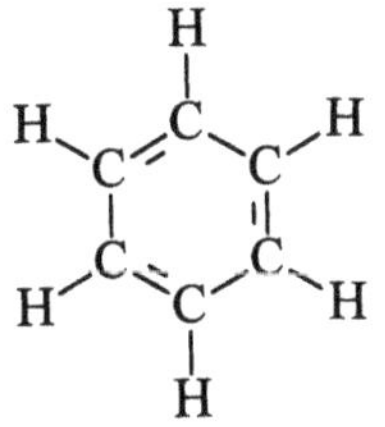

Figure 1.5. The benzene ring.

skeletons, or basic structures, fall into two major categories: aromatic and aliphatic. All aromatic organic compounds contain a structure called a benzene ring, shown in Figure 1.5.

The bonds between the carbon atoms alternate between a single and a double bond. This unique structure is what separates the two major categories of organic compounds. All organics that do not contain a benzene ring are aliphatic compounds. Explosives fall into both these categories.

The aromatic and aliphatic terms thus refer to the basic carbon skeleton of the molecule. This carbon skeleton and any attached hydrogen atoms constitute the fuel portion of the explosive. The oxidizer portion is in the form of attached subgroups containing oxygen. Table 1.2 contains some of the more common subgroups, both oxidizer and fuel types.

1.2.1.1 Aromatic Explosives

As stated above, aromatic explosives contain a benzene ring. In all these explosives the benzene is nitrated, usually with three nitro groups. They are all solid crystalline materials. Their melting points vary from 65 to 260°C. This nitrated benzene, trinitrobenzene (TNB), is the basic building block and is shown in Figure 1.6.

Three types of aromatics are discussed here. In the first set, the TNB has one additional molecular subgroup substituted for a hydrogen atom on the ring. In

Table 1.2 Molecular Subgroups

Name	Elemental Formula
Nitrate	NO_3
Nitro	NO_2
Azide	N_3
Ammonium	NH_4
Amino	NH_2
Hydroxyl	OH
Methyl	CH_3
Ethyl	C_2H_5

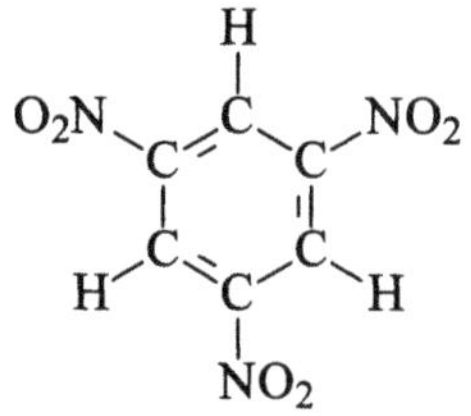

H
O_2N C NO_2
C C
C C
H C H
NO_2

Figure 1.6. The trinitrobenzene molecule.

the second set, the aromatics have two or more molecular subgroups substituted for hydrogen atoms on the ring. The third set has two or more nitrated benzene rings in the same molecule.

Singly Substituted TNB: Trinitrotoluene (TNT) is the most common aromatic explosive. Figure 1.7 depicts the molecule structure, where we see that the hydrogen atom at the top of the benzene ring has been replaced by or substituted with a methyl group. More TNT is produced than any other military explosive. Because of its low melting point and excellent safe handling characteristics, it is used as a blending stock to which many other explosives are added. Its melting point is about 81°C, and it is usually melt cast into the forms in which it is used. It is stable, insensitive to shock, and nontoxic. It is made from toluene, a petroleum byproduct. The nitro groups are added to it in steps, reacting the toluene in a mixture of concentrated nitric acid with some sulfuric acid added. If the methyl group, —CH_3, is removed from the TNT and replaced with other groups, we have a large family of singly substituted aromatic explosives, as shown in Figure 1.8.

Multiply Substituted TNB: This series of explosives, depicted in Figures 1.9 through 1.15, is based on TNB but has more than one additional group attached. The trinitroxylene molecule (TNX) forms a pale yellowish needle crystal with a negative oxygen balance (89.57%). This material is difficult to manufacture; consequently, it is not a popular explosive.

The trinitroresorcinol molecule (styphnic acid). This explosive forms a yellow-brown to red-brown crystal with a negative oxygen balance (−35.9%). It is

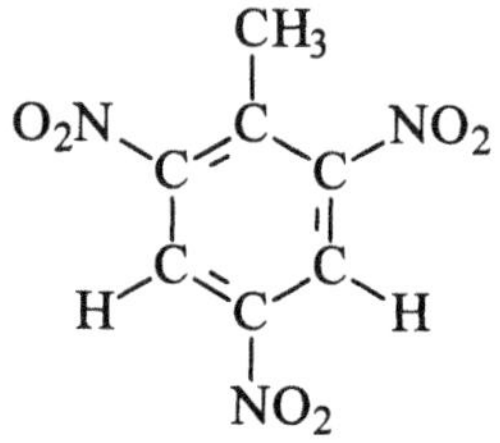

CH_3
O_2N C NO_2
C C
C C
H C H
NO_2

Figure 1.7. The trinitrotoluene molecule.

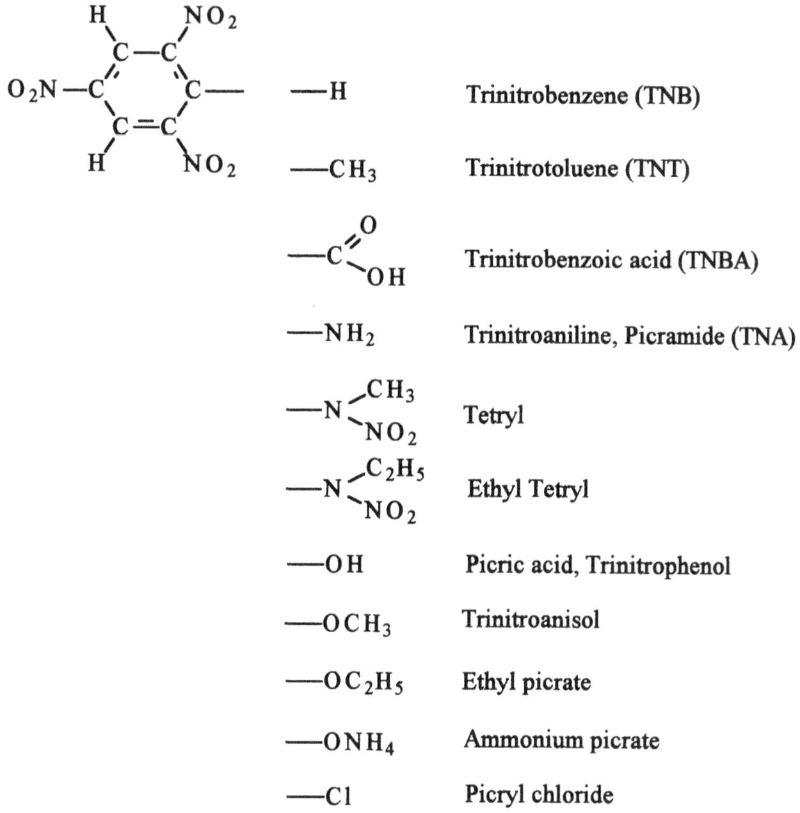

Substituent	Compound
—H	Trinitrobenzene (TNB)
—CH_3	Trinitrotoluene (TNT)
—C(=O)OH	Trinitrobenzoic acid (TNBA)
—NH_2	Trinitroaniline, Picramide (TNA)
—N(CH_3)(NO_2)	Tetryl
—N(C_2H_5)(NO_2)	Ethyl Tetryl
—OH	Picric acid, Trinitrophenol
—OCH_3	Trinitroanisol
—OC_2H_5	Ethyl picrate
—ONH_4	Ammonium picrate
—Cl	Picryl chloride

Figure 1.8. A family of explosives based on various monosubstitutions of a single hydrogen on trinitrobenzene.

a relatively weak explosive and not very popular, but its lead salt (lead styphnate) is used as an initiating explosive. The lead styphnate molecule is a sensitive explosive used extensively as an initiating material. It has replaced mercury fulminate in gun primers and is used in other initiating explosive devices. It is often mixed with lead azide, forming a detonator first fire charge.

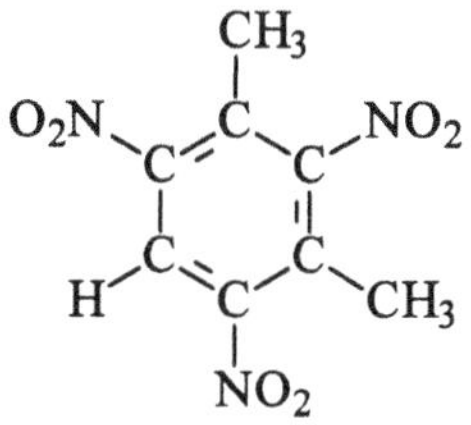

Figure 1.9. Trinitroxylene (TNX).

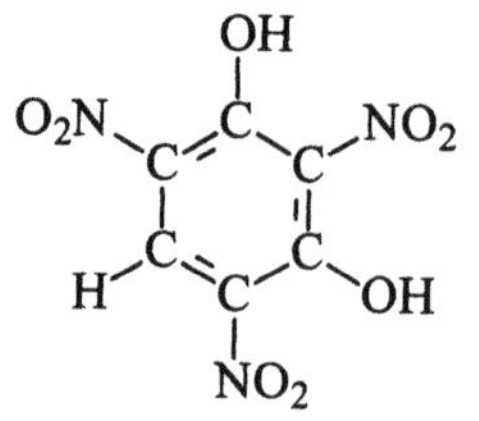

Figure 1.10. Trinitroresorcinol (styphnic acid).

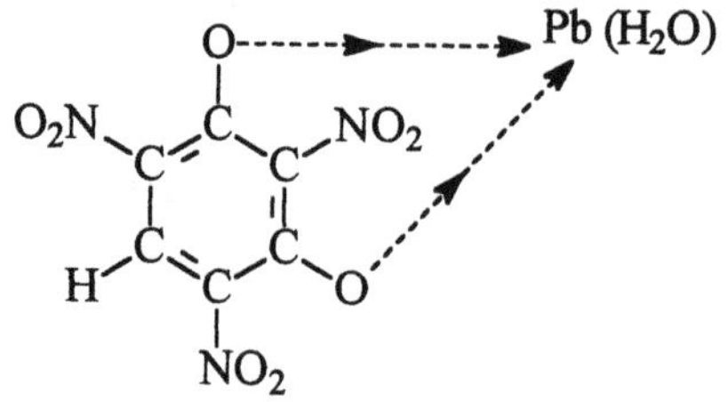

Figure 1.11. Lead styphnate.

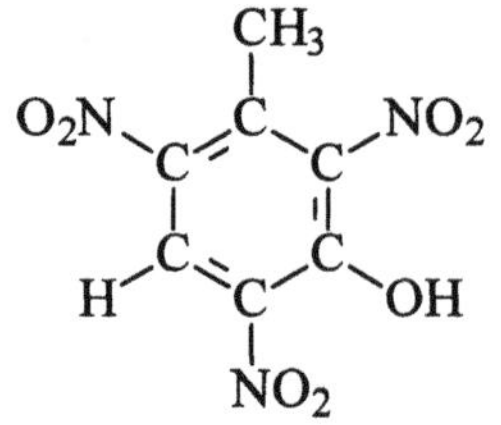

Figure 1.12. Trinitrocresol (TNC).

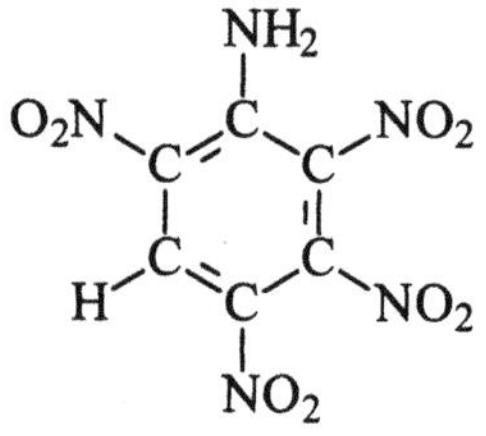

Figure 1.13. Tetranitroaniline (TNA).

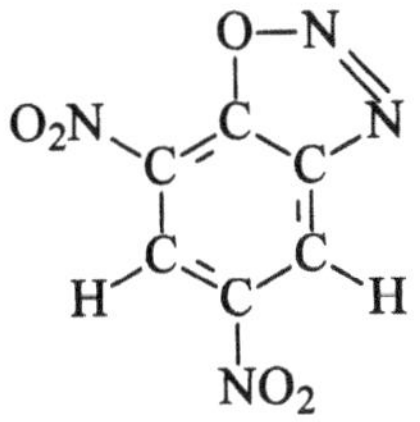

Figure 1.14. Diazodinitrophenol (DINO or DDNP).

The trinitrocresol molecule (TNC) was used extensively during World War I by the Germans as a bomb and hand grenade filler. It has a melt temperature of approximately 85°C, and in the melt stage, it was mixed with picric acid and poured directly into the bombs.

The tetranitroaniline molecule (TNA) forms pale yellow crystals with a negative oxygen balance (−32.2%). TNA mixed with 5% paraffin was used in shells, bombs, and mines as a booster charge, but later discontinued because of its water solubility.

The diazodinitrophenol molecule (DINO, DDNP) forms a reddish yellow, amorphous powder with a large negative oxygen balance (−60.9%). It has been used as an initiating explosive and is slightly more powerful than mercury fulminate and less powerful than lead azide.

The triaminotrinitrobenzene molecule (TATB) is very stable to temperatures up to 300°C and is extremely insensitive to friction and impact. It is commonly known as the "insensitive" explosive. This explosive should never be in direct contact with copper and other heavy metals.

Multiple TNB Rings: Aromatic explosive molecules can contain two or more nitrated benzene rings in the same molecule. The more important of these multiple TNB ring explosives are shown in Figures 1.16 through 1.19. The hexanitroazobenzene (HNAB) molecule is a thermally stable explosive with an orange to red crystal. It has a negative oxygen balance (−49.7%) and is used in some special military explosive trains.

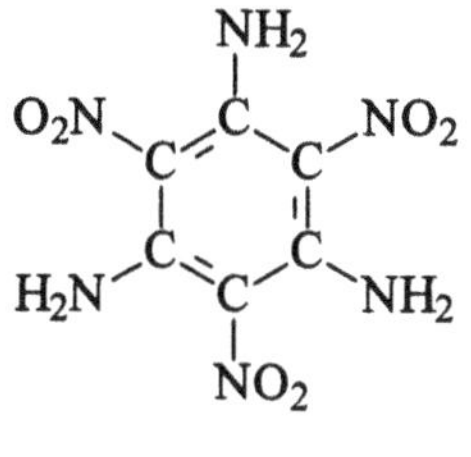

Figure 1.15. Triaminotrinitrobenzene (TATB).

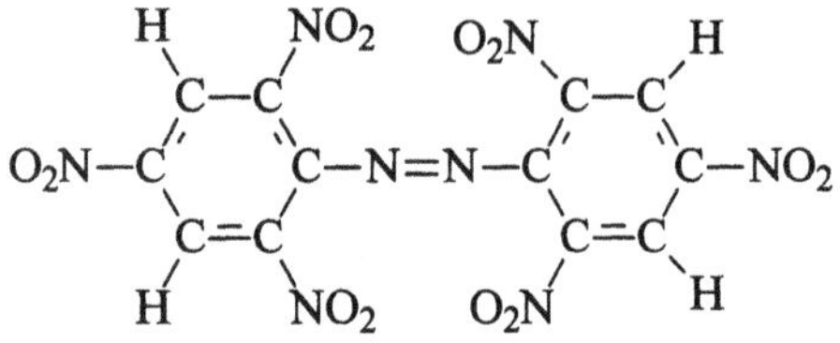

Figure 1.16. Hexanitroazobenzene (HNAB).

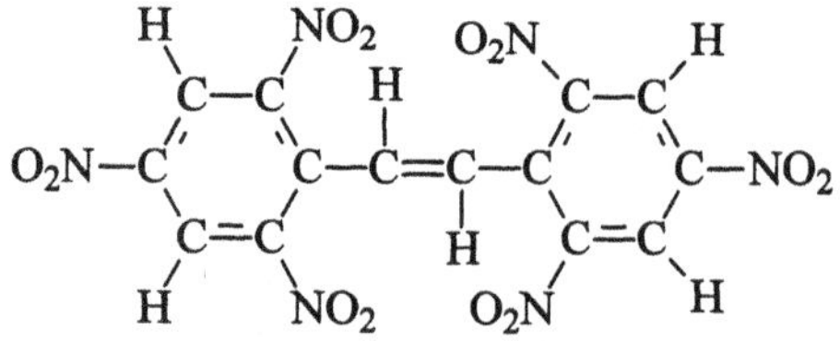

Figure 1.17. Hexanitrostilbene (HNS).

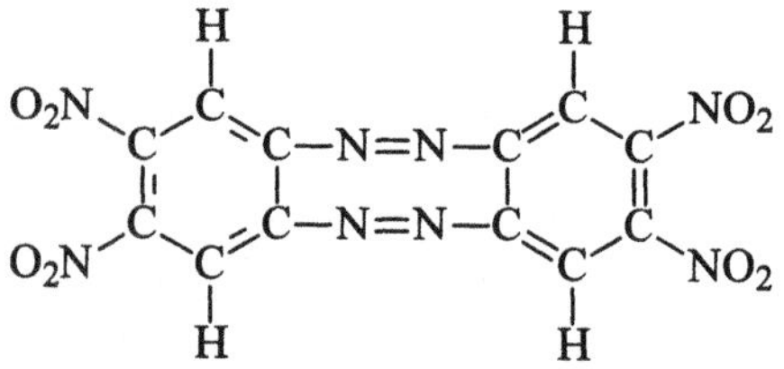

Figure 1.18. Tetranitrodibenzotetrazapentalene (TACOT).

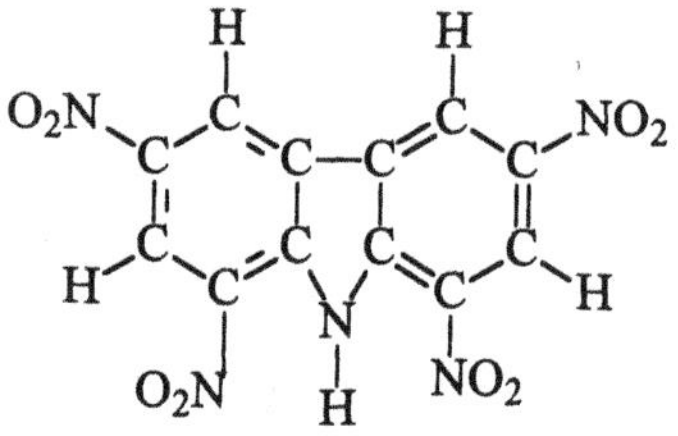

Figure 1.19. Tetranitrocarbazole (TNC).

Hexanitrostilbene (HNS) is also thermally stable and is used by the U.S. Department of Energy and the U.S. Department of Defense as well as in many space applications. This yellow crystal explosive can be pressed into structurally sound pellets and is used extensively in heat-resistant detonating trains in weapon systems. It is not yet common in commercial products.

Tetranitrodibenzotetrazapentalene (TACOT) is a commercial, thermally stable explosive made by E. I. Dupont, Inc. It is exceptionally stable at high temperatures, which range from 350°C for 10 minutes to 275°C for 4 weeks.

The tetranitrocarbazole molecule (TNC): Notice there are two explosives called TNC. Fortunately, neither is widely used. This explosive is insoluble in water, which makes it a good candidate for components used in and around water.

1.2.1.2 Aliphatic Explosives

As mentioned earlier, the aliphatic organic compounds do not contain a benzene ring. These explosives were the first modern explosives to be mass produced. The most common aliphatic explosives are the nitrate esters.

Aliphatic Nitrate Esters: The nitrate esters comprise a large number of short-chain explosives that contain the nitrate group, —ONO_2. A host of very similar nitrate esters, including some only partially nitrated higher-order alcohols, are all very similar in chemical nature and physical properties. Some are oily liquids, insoluble in water, with moderate-to-high vapor pressures. They are, for the most part, very sensitive to impact. They are used in dynamites and in double-base propellants. Since they gelatinize nitrocellulose, they are present in the propellants, not only for energy, but also to moderate the physical properties of the mix.

These liquid explosives are very sensitive because of their vapor pressure, which causes microscopic bubbles to form in the liquid. The bubbles act as hot spots or initiating sites in the material when it is subjected to shocks.

Nitroglycerine was the first, and is still one of the most widely produced, nitrate ester. Used in dynamites, nitroglycerine is absorbed in fine wood meal or other powdered absorbents. This prevents the microbubbles from forming and stabilizes the liquid. The nitroglycerine is also thickened or gelatinized by addition of a small percentage of nitrocellulose. This helps prevent the liquid from "weeping" or settling out of the absorbent material. Because settling does occur, boxes of stored nongelled dynamites are turned over at regular intervals to reverse the settling flow. Nitroglycerine (Figure 1.20), like other aliphatic nitrate esters, is made by nitrating the alcohol groups with nitric acid. Nitroglycerine is made by nitrating glycerine, a polyalcohol.

Figures 1.21 and 1.22 show other short-chain nitrate ester explosives similar to nitroglycerine.

Pentaerythritol tetranitrate (PETN) is another major aliphatic nitrate ester. See Figure 1.23. This white, powdered explosive is the major constituent in detonating cords. It is also used in many mixed and cast explosives, and as a pure

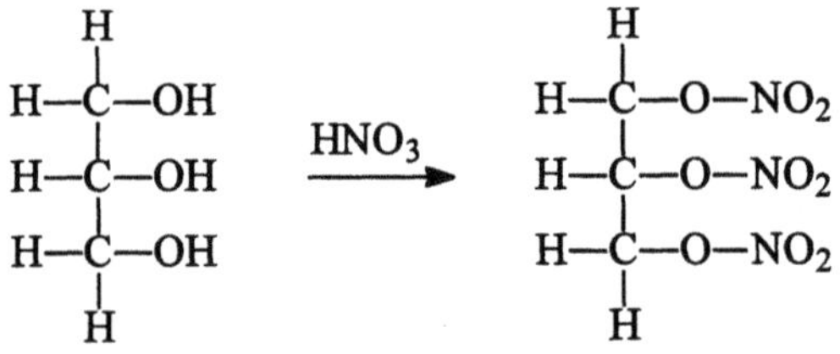

$$
\begin{array}{c}
\mathrm{H} \\
| \\
\mathrm{H{-}C{-}OH} \\
| \\
\mathrm{H{-}C{-}OH} \\
| \\
\mathrm{H{-}C{-}OH} \\
| \\
\mathrm{H}
\end{array}
\xrightarrow{HNO_3}
\begin{array}{c}
\mathrm{H} \\
| \\
\mathrm{H{-}C{-}O{-}NO_2} \\
| \\
\mathrm{H{-}C{-}O{-}NO_2} \\
| \\
\mathrm{H{-}C{-}O{-}NO_2} \\
| \\
\mathrm{H}
\end{array}
$$

Glycerine **Nitroglycerine**

Figure 1.20. Nitroglycerine is made by nitrating glycerine, a polyalcohol.

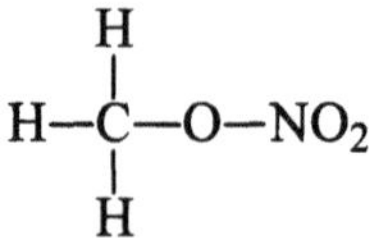

$$
\begin{array}{c}
\mathrm{H} \\
| \\
\mathrm{H{-}C{-}O{-}NO_2} \\
| \\
\mathrm{H}
\end{array}
$$

Figure 1.21. The methylnitrate molecule; note that this is not nitromethane.

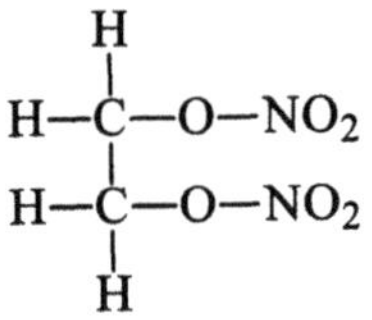

$$
\begin{array}{c}
\mathrm{H} \\
| \\
\mathrm{H{-}C{-}O{-}NO_2} \\
| \\
\mathrm{H{-}C{-}O{-}NO_2} \\
| \\
\mathrm{H}
\end{array}
$$

Figure 1.22. Nitroglycol.

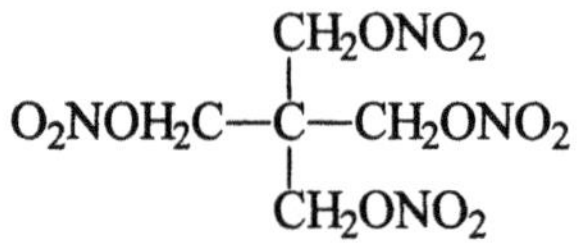

$$
\begin{array}{c}
\mathrm{CH_2ONO_2} \\
| \\
\mathrm{O_2NOH_2C{-}C{-}CH_2ONO_2} \\
| \\
\mathrm{CH_2ONO_2}
\end{array}
$$

Figure 1.23. Pentaerythritoltetranitrate (PETN).

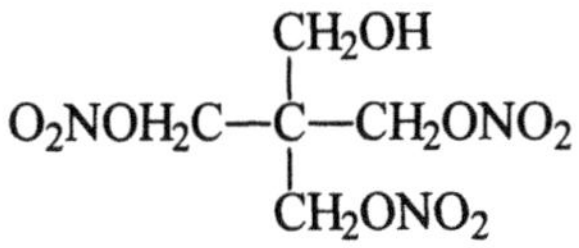

$$
\begin{array}{c}
\mathrm{CH_2OH} \\
| \\
\mathrm{O_2NOH_2C{-}C{-}CH_2ONO_2} \\
| \\
\mathrm{CH_2ONO_2}
\end{array}
$$

Figure 1.24. Pentaerythritoltrinitrate (PETRIN).

```
      H
      |
H—C—NH—NO2
      |
H—C—NH—NO2
      |
      H
```

Figure 1.25. Ethylenedinitramine (EDNA).

powder in blasting cap output charges and in initial pressings of exploding bridgewire detonators. The major drawback of PETN is that it is not sufficiently stable above 71°C, and although classed as a secondary explosive, it is one of the most impact sensitive of the secondary explosives. However, because of its shock sensitivity, it is used extensively in exploding bridgewire detonators. Pentaerythritoltrinitrate (PETRIN) (Figure 1.24), a close cousin to PETN, is used in many European explosives and in some propellant formulations.

Aliphatic Nitramines: Nitramines, —N—NO_2, are another group of aliphatic explosives. They are somewhat similar in structure to the nitrate esters. However, these explosives melt at much higher temperatures. One in wide use is ethylene dinitramine (EDNA) (Figure 1.25), also called halite, which melts at about 176°C.

Another commonly used aliphatic nitramine is nitroguanidine (NQ) (Figure 1.26). Nitroguanidine is used not only as an explosive, but also as a major constituent in triple-base propellants. In one of its crystal habits, it has a unique featherlike structure, and because of this can be pressed stably at very low densities. This property makes it a valuable tool where extremely low detonation velocity and detonation pressure are needed.

Two of the nitramines, RDX and HMX, are very important due to their relatively high thermal stability, high density, and high detonation velocity. See Figures 1.27 and 1.28. These two compounds are ring compounds but are not aromatic. The rings are not composed of six carbon atoms with alternating double bonds. These are cyclic aliphatic structures that include nitrogen atoms alternating with the carbon atoms. These two stable, high-velocity explosives are used as pressed powders, in castable mixes with TNT, and in plastic-bonded and plasticized forms.

Nitro-Aliphatics: The aliphatic nitro compounds contain the nitro, —NO_2, bonded directly to carbon. It is not bonded through oxygen like the nitrates or

```
          NH2
         /
HN=C
         \
          NH—NO2
```

Figure 1.26. Nitroguanidine (NQ).

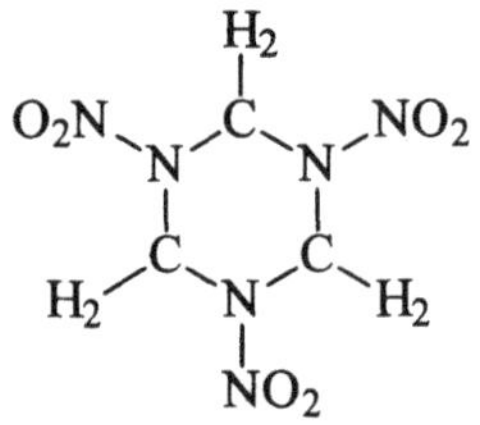

Figure 1.27. Hexahydrotrinitrotriazine (RDX, also cyclonite or hexogen).

through nitrogen, like the nitramines. The simplest of the nitro-aliphatics is nitromethane, CH_3NO_2. Nitromethane (NM) is a liquid. Nitromethane is not as sensitive or as toxic as the liquid nitrate esters, and is used either by itself or in mixtures with other explosives or mixed with an oxidizer as in binary explosives. NM is used far more often as a solvent or as a high-performance fuel for internal combustion engines than it is as an explosive.

1.2.2 Pure Compounds: Inorganic

Inorganic materials are not based upon a carbon skeleton. Most inorganic explosives are mixtures of fuels and oxidizers. In this section we will only consider several pure or single-compound inorganic explosive molecules.

Ammonium Nitrate (NH_4NO_3): On the basis of commercially produced tonnage, this is certainly the most common explosive. See Figure 1.29. By itself, it is a rather poor explosive, being very overoxidized (OB = +20%) and extremely difficult to initiate. It is used safely in its pure state as a fertilizer in multitonnage quantities. When mixed with other explosives or with fuels in the form of powder, slurry, or gel mixtures, it becomes an effective an inexpensive commercial explosive for use in blasting.

Fulminates: These pure inorganic explosives are all primary, or initiating, explosives. Mercury fulminate, $Hg(0NC)_2$, shown in Figure 1.30, was one of the first pure inorganic explosives to be used. This extremely sensitive crystalline material was used in early electric blasting caps and also in percussion primers

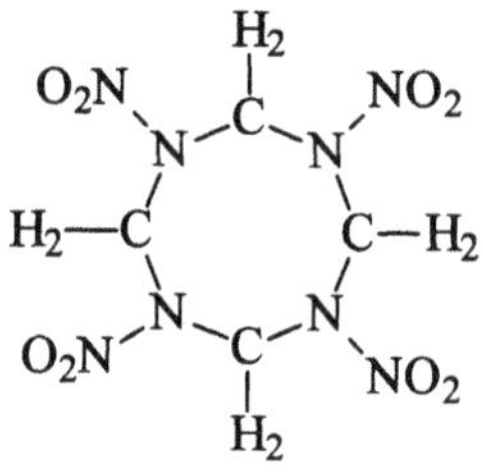

Figure 1.28. Octahydrotetranitrotetrazine (HMX, also octogen).

$$(NH_4)^+ (NO_3)^-$$

Figure 1.29. Ammonium nitrate (AN).

for small arms. Because of its poor storage characteristics and corrosiveness when exposed to even small amounts of moisture, it is seldom, if ever, used anymore.

Silver fulminate, AgONC, is considerably more sensitive than mercury fulminate. This explosive is seldom intentionally made. However, it is occasionally accidentally produced during the liquid silvering process in making mirrors. It forms when alcohol comes in contact with acidic silver nitrate solutions. A number of laboratories have experienced minor explosions from inadvertent synthesis of silver fulminate.

Silver Acetylide (C_2Ag_2): This is another explosive that is sometimes made accidentally. It is formed by bubbling acetylene through a solution of silver nitrate. Silver acetylide is so sensitive that it sometimes detonates as the crystals are forming or drying.

Azides: These are salts of hydrazoic acids (HN_3). Alkali azides are the most important intermediates in the production of lead azide, the more common of the two explosives in the azide family.

Lead azide (PbN_6) is extremely sensitive to sparks, friction, and impact. It is the major initiating explosive used in virtually all blasting caps and most other hot-wire initiated detonators. It has excellent storage characteristics and can be pressed quite easily if it has been phlegmatized (lubricated or coated). It has excellent thermal characteristics, being stable up to almost 500°F. In its absolutely pure form, it is too sensitive to be handled because it retains electrostatic charges. Phlegmatization or co-precipitation with dextrin cures this major drawback.

Another explosive in the azide family, silver azide (AgN_3), is somewhat sensitive to light and therefore does not store well in its unpressed powdered form. It is stable to slightly higher temperatures than lead azide, and once in a light-tight pressing, will store well.

Copper azide is an extremely sensitive explosive and can be formed unintentionally when lead azide comes in contact with copper in the presence of moisture.

Sodium azide (NaN_3) is not a detonatable explosive but is an excellent propellant. It is the major constituent in gas generators for automobile safety air bags.

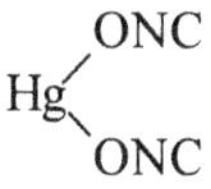

$$Hg\begin{matrix} \diagup ONC \\ \diagdown ONC \end{matrix}$$

Figure 1.30. Mercury fulminate.

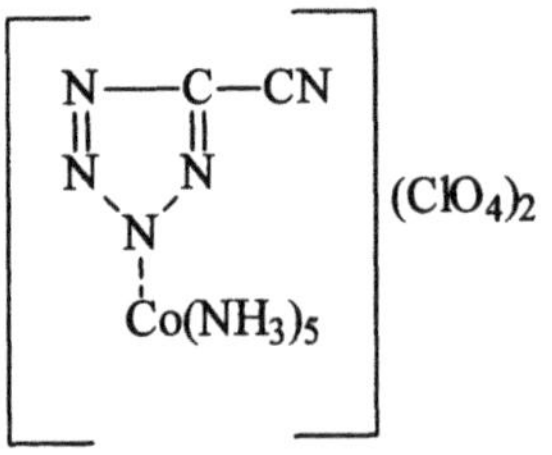

N—C—CN
N N
N
$Co(NH_3)_5$
$(ClO_4)_2$

Figure 1.31. Cyanotetrazolatopentaamine cobalt III perchlorate (CP).

Cyanotetrazolatopentaammine cobalt III perchlorate (CP): This unique explosive will reliably transfer from burning to a detonation (DDT) in a small diameter. Therefore, it can be used in detonators using a hot-wire input.

Unlike a true primary explosive, it is less sensitive to shock and electrostatic discharge. This explosive is used mainly by the U.S. Department of Energy for special detonators. It is rarely used in the commercial market. Figure 1.31 shows the CP molecule.

1.3 Use Forms of Explosives

Although the pure explosive compounds described are used in their pure form as liquids, pressed powders, or in some cases, such as with TNT, as castings, the majority of uses for explosives require mechanical properties that the pure materials do not have. To change or alter the mechanical properties, as well as some of the thermal, output, or sensitivity properties, the pure explosives are blended with other explosives and inert materials. The mixtures can then be worked in various ways to form specific explosive products. These product forms are:

Pressings,
Plastic bonded (machinable),
Putties,
Extrudables,
Slurries and emulsions,
Binary or two component (liquids and solids),
Blasting gelatin,
Liquids, and
Gases.

1.3.1 Pressings

Many of the explosive materials have crystal forms that are not amenable to pressing operations. The pressed pellets do not hold together well and can easily flake apart. Some powders are extremely sensitive to initiation by electrostatic buildup or friction during pressing. Others simply will not flow well enough into pressing dies or molds. To alleviate these problems, various additives are blended with the explosives. Molding lubricants are generally either graphite or stearates. Phlegmatizing agents such as petroleum jelly and mineral oil are used. Colorants and taggants, as well as antistatic agents, are also added to some pressing mixes.

There are two basic types of pressing operations, constant volume and constant force. In the first, the ram of the press, under very high force, is pressed to a mechanical stop in the die. Pellets pressed in this manner have very precise physical dimensions or constant volume, but the density of the pellet can vary according to how accurately the powder charge was weighed. In the second method, the die has no mechanical stop. The ram is driven by a precise force onto the powder and presses the pellet to a rather precise or controlled density. However, in this case the length of the pellet can vary and depends upon the accuracy of the charge weight.

1.3.2 Castings

All the modern castable formulations are based upon mixtures of relatively higher melting crystalline explosives with molten TNT. Since TNT has such a negative oxygen balance, oxidizers such as nitrates and positive oxygen balance explosives are often added to it. Many of the castable explosives are also machinable. Their machinability, however, is not as good as that of plastic-bonded explosives. Table 1.3 shows many of the TNT-based castables in use.

1.3.3 Plastic Bonded (PBX)

These are powdered explosives to which plastic or polymeric binders have been added. The binder is usually precipitated out of solution in the preparation process such that it coats the explosive crystals. Agglomerates of these coated crystals form pressing "beads." The beads are then either die pressed or isostatically pressed at temperatures as high as 120°C. Pressures from 10,000 to 20,000 psi then produce pellets, or billets, with densities that may approach 99% of the theoretical maximum density. The billets thus produced have good mechanical strength and can be machined to very close tolerances.

Table 1.4 lists various PBXs that have been developed by the U.S. Department of Energy weapons laboratories. Table 1.5 shows the PBX ingredients and their functions.

1.3.4 Putties

These explosives are mixtures of finely powdered RDX and plasticizers. The mixture is puttylike and can be molded by hand to any desirable shape. Like

Table 1.3 TNT-Based Castable Explosives

Name	TNT	Aluminum	Ammonium Nirate	Ammonium Picrate	Barium Nitrate	Boric Acid	Calcium Chloride	EDNA	HMX	Lead Nitrate	PETN	Sodium Picrate	RDX	Tetryl	Wax
Amatex	x		x												
Amatol	60		40												
Amatol	50		50												
Amatol	20		80												
Ammonal	67	11	22												
Baronal	35	15			50										
Baratol	33				67										
Baratol	24				76										
Boracitol	40					60									
Comp B	36												63		1
Comp B-2	40												55		5
Comp B-3	40												60		
Cyclotol	50												50		
Cyclotol	35												65		
Cyclotol	30												70		
Cyclotol	25												75		
DBX (Minex)	40	18	21										21		
Ednatol	45							55							
H-6	30	20					0.5						45		5
HBX-1	38	17					0.5						40		5
HBX-3	29	35					0.5						31		5
HTA-3	29	22							49						
Minol-2	40	20	40												
Octol	23								77						
Octol	25								75						
Octol	30								70						
Pentolite	50										50				
Pentolite	90										10				
Picratol	48			52											
Plumbatol	30									70					
PTX-1	20												30	50	
PTX-2	31										27		42		
Sodatol	50											50			
Tetrytol	20													80	
Tetrytol	25													75	
Tetrytol	30													70	
Tetrytol	35													65	
Tritonal	80	20													
Torpex	40.5	18											40.5		1

Table 1.4 Plastic-Bonded Explosives

Name	Explosive Ingredients (%)	Binder Ingredients	Color
LX-14-0	HMX 95.5	Estane 5702-F1	Violet spots on white
LX-10-0	" 95.0	Viton A	Blue-green spots on white
LX-10-1	" 94.5	Viton A	Blue-green spots on white
PBX9501	" 95.0	Estane, BDNPA-F	White
PBX9404	" 94.0	NC, CEF	Blue or white
LX-09-1	" 93.3	BDNPA-F, FEFO	Purple
LX-09-0	" 93.0	BDNPA-F, FEFO	Purple
LX-07-2	" 90.0	Viton A	Orange
PBX9011	" 90.0	Estane 5702-F1	Off-white
LX-04-1	" 85.0	Viton A	Yellow
LX-11-0	" 80.0	Viton A	White
LX-15	HNS-1 95.0	Kel-F 800	Beige
LX-16	PETN 96.0	FPC 461	White
PBX9604	RDX 96.0	Kel-F 800	—
PBX9407	" 94.0	FPC 461	White or black
PBX9205	" 92.0	Polystyrene, DPO	White
PBX9007	" 90.0	Polystyrene, DOP, rosin	White or mottled grey
PBX9010	" 90.0	Kel-F 3700	White
PBX9502	TATB 95.0	Kel-F 800	Yellow
LX-17-0	" 92.5	Kel-F 800	Yellow
PBX9503	80TATB/15HMX	Kel-F 800	Purple

Table 1.5 PBX Binder and Plasticizer Ingredients and Functions

Name	Material	Function
Estane 5702-F1	Polyurethane solution system	Binder
Viton A	Vinylidine fluoride (60%)/hexafluoropropylene (40%) copolymer	Binder
NC	Nitrocellulose (12.0% N)	Reactive binder
CEF	Tris(beta)-chloroethylphosphate	Plasticizer
BDNPA-F	Bis(2,2-dinitropropyl)acetal/bis(2,2-dinitropropyl) formal (50/50 wt %)	Reactive plasticizer
Kel-F 800 and 3700	Chlorotrifluoroethylene/vinylidine fluoride copolymers	Binder
FPC 461	Vinyl chloride/chlorotrifluoroethylene copolymer, 1.5:1	Binder
Polystyrene	Polystyrene	Binder
DOP	Di(2-ethylhexyl) phthalate	Plasticizer

Table 1.6 Composition C-4

Component	Weight (%)
RDX	91.0
Di(2-ethylhexyl sebacate	5.3
Polyisobutylene	2.1
20 weight motor oil	1.6

modeling clay, it retains its shape unsupported after molding. Although many different putty compositions have been made, only one is now prevalent in the United States. That is the military explosive Composition C-4. See Table 1.6 for the formulation. In the United Kingdom there is a similar military putty explosive called P-4. A series of commercial putty explosives are produced in Czechoslovakia at the huge explosives works in the city of Semptin. These use various combinations of PETN and RDX with different binders. The products have specific designations depending upon the particular composition, but as a group are referred to as Semtex.

1.3.5 Rubberized

Mixtures of RDX or PETN with rubber-type polymers and plasticizers can be rolled into rubbery gasketlike sheets. The sheets maintain their dimensional stability and are very easy to handle. They can be cut to specified shapes and glued to a desired surface. The commercial product is manufactured by Dupont under the trade name "Detasheet," and by North American Explosives (NAX) under the name "Prima-Sheet." Sheets come in varying thicknesses and weights (Table 1.7). Although several types (varying percentages of high explosives) were formerly made, only the "Detasheet C" is currently available from Dupont. It consists of 63% PETN, 8% nitrocellulose (12.34% nitration), and 29% acetyltributylcitrate (ATBC). Both an RDX and a PETN version are available from NAX: PS2000 and PS1000, respectively.

Another version of Detasheet, called Deta Flex, is manufactured for the military. It contains the same plasticizers, but has approximately 70% RDX instead of the 63% PETN. It comes only in one-quarter-inch thickness and is olive colored.

A version of this explosive formulated by the U.S. Department of Energy is called LX-02-1. This material is colored blue and contains 73.5% PETN, 17.6% butylrubber, 6.9% ATBC, and 2% Cab-o-sil.

1.3.6 Extrudables

PETN or RDX mixed with uncured silicone rubber resin at 80% PETN and 20% rubber forms a thick viscous material that can be cold extruded under moderate

Table 1.7 Sheet Explosive Designations, Sizes, and Weights

Designation	Thickness (in.)	Weight (g/in.^{2})
C-1	0.04	1
C-2	0.08	2
C-3	0.12	3
C-4	0.16	4
C-6	0.24	6
C-10	0.40	10
1000	0.042	1
1000	0.062	1.5
1000	0.083	2
1000	0.125	3
1000	0.250	6
1000	0.333	8
2000	0.083	2
2000	0.125	3
2000	0.208	5
2000	0.250	6
2000	0.300	7

pressures (less than 100 psi). After extrusion into holes, molds, or channels, the temperature is raised and the resin polymerizes and cures, leaving a tough rubberlike material. Three U.S. Department of Energy specifications exist for this type of explosive: LX-13 (green), XTX-8003 (white), and XTX 8004 (white). All three are 80/20 mixes of the above; they differ in the type and particle size of the explosive used. LX-13 and XTX-8003 use PETN, and XTX-8004 uses RDX explosive. A similar extrudable explosive developed by the U.S. Navy is included in the PBXN series.

1.3.7 Slurries and Emulsions

These explosives are probably produced in greater quantities than any other. They were introduced starting in the late 1950s. This mainstay of commercial blasting is a thickened supersaturated solution of ammonium nitrate in water. As stated earlier, ammonium nitrate, NH_4NO_3, has a very high positive oxygen balance (+20%). The most common fuel used in slurries is aluminum powder. However, many other fuels are used also, among them water-soluble fuels such as glycol and alcohols. The slurries are extremely insensitive to initiation; so many of them are sensitized by addition of other powdered explosives such as PETN and TNT. They are also sensitized by the addition of glass microballoons. The microballoons provide a source of tiny bubbles, which are required for the stable propagation of the detonation wave through this material. Slurries are thickened with gelatins, guar gums, and water-soluble polymerizable plastics. Sensitized, heavily gelled slurries are packaged in cartridges, like dynamites.

Slurries can also be brought, unmixed, to the use site, where they are pumped from tanker trucks into mixing valves and from there are pumped directly into the blast pattern boreholes. This method uses most of the tonnage of slurries. The amount of slurries used in 1980 in the United States alone exceeded 4 billion pounds. As a comparison, at the height of the war in Viet Nam, the military was using about a half-billion pounds of explosives annually.

Emulsions are essentially slurries turned inside out. Whereas a slurry has fuel dispersed in an aqueous matrix, emulsions consist of tiny droplets of aqueous oxidizers dispersed in an oil matrix. The sensitivity and to some extent the detonation properties are adjusted by the addition of various sizes of either glass or plastic microballoons. Emulsions have the advantage over slurries in some blasting applications in that they are less prone to problems caused by the intrusion of water.

1.3.8 Binary or Two Component (Liquids and Solids)

In this concept, two nonexplosive components are mixed together just before use to form a cap-sensitive high explosive.

Astrolite is a two-part system consisting of a white powder (proprietary, but assumed to be ammonium nitrate) and slightly aqueous hydrazine. These components, when mixed, form a thin, waterlike, clear liquid that is cap sensitive and detonates at very small diameters.

XPLO Corporation produces a liquid two-component system called Marine Pac that consists of part A, a yellow liquid (nitroparafin), and part B, a red liquid (secondary amine/aliphatic and alicyclic polyamines).

Kine-Pak and Kine-Stick consist of a mixture of ammonium nitrate and nitromethane, and are manufactured by the Atlas Powder Company.

1.3.9 Dynamites, Blasting Gelatin, and Blasting Agents

Dynamites are cap-sensitive high explosives usually containing nitroglycerin (straight dynamite), or nitroglycerin gelatinized with nitrocellulose (gelatin dynamite), or either of the above two mixed with ammonium nitrate (extra and extra-gelatin dynamites). These traditional types of dynamite are now supplemented with a number of non-nitroglycerin containing explosives mainly based on ammonium nitrate mixtures that have been made cap sensitive. In the nitroglycerin-based dynamites the liquid explosive is absorbed into a bed of wood meal or similar absorbing material. This is done to prevent the formation of tiny bubbles of vapor in the nitroglycerin. These sometimes microscopic bubbles are the main reason that nitroglycerin is so sensitive to initiation by shock or impact. Some dynamites have flame-suppressant chemicals added to them such as halogen salts. This is done for dynamites used in mines where explosive vapors and dusts may be present. The flame-suppressant salts help prevent ignition or explosion of these vapors. Such dynamites are called "permitted" dynamite. Dynamites are usually packaged as a paper tube or roll filled with the explosive. These "sticks" range in weight from as little as an eighth of a pound up to a full pound or more each.

Blasting gelatin is made by gelatinizing nitroglycerin with nitrocellulose or guncotton. The gel is adjusted to approximately 93% nitroglycerin. This material is the standard against which dynamites are compared. Thus, when a dynamite is designated as, for example, 60% dynamite, this means that its output is 60% of that of standard blasting gelatin in a ballistic mortar test. Blasting gelatin is rarely used in commercial blasting because it is too powerful.

Blasting agents are insensitive explosive mixtures usually made with ammonium nitrate as the oxidizer and petroleum oils as the fuel. Blasting agents cannot be initiated with a blasting cap and therefore are handled differently than dynamites. Many of the slurry and emulsion explosives mentioned earlier are classed as blasting agents.

1.3.10 Liquids

The major liquid explosive is nitromethane and various additives. Nitromethane was used in large quantities in underground blast studies at the Nevada Test Site. Some use has been made in attempting to stimulate oil and gas well production with detonations of nitromethane and additives.

1.3.11 Gases

Other than detonation of fuel/air mixtures in some experimental military weapons, very little application has been made, or is anticipated, of this physical phase of explosive materials.

1.4 Propellants

Propellants appear, chemically, to be quite similar to explosives. In fact, some explosives are also used in and as propellants. However, in their use mode, propellants burn, but do not detonate. Propellants give off gas and heat, but do not react faster than the bulk sound speed in the propellant material. As a matter of fact, typical gun propellants burn quite slowly. Burning velocities at one atmosphere pressure are in the range of 0.1 to 1 mm/s (that is around 10 million times slower than detonation velocities!). Like explosives, typical gas output is in the neighborhood of 1 liter of gas (at standard temperature and pressure) per gram of propellant, and heat output is around 1 kcal/g.

Chemically, propellants are divided into four major categories. They are:

- Single base,
- Double base,
- Triple base, and
- Composite.

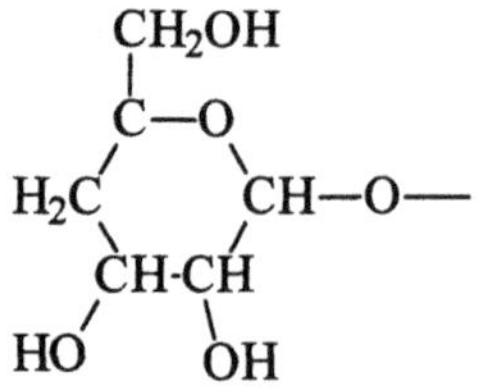

```
      CH2OH
        |
        C—O
       /   \
   H2C      CH—O—
       \   /
       CH-CH
      /     \
    HO       OH
```

Figure 1.32. Cellulose monomer.

1.4.1 Single-Base Propellants

Single-base propellants are essentially pure nitrocellulose. This material is made from cellulose, a natural fibrous polymer found in plants. Common sources of cellulose in industry are cotton and wood pulp. The empirical formula of cellulose is $(C_6H_{10}O_5)_n$. The n signifies the number of $C_6H_{10}O_5$ groups, or monomers, that are connected in the long polymer chain. Figure 1.32 shows the structural formula of the monomer. Large numbers of these monomers are connected to each other end to end to form a plastic or polymer material. See Figure 1.33. In general, the longer the chain, the stiffer the material.

The hydroxyl groups (—OH) in the cellulose are nitrated by reaction with a mixture of nitric and sulfuric acids. The hydroxyl on the side-chain carbon is nitrated first. After all the side-chain groups are nitrated, the other two hydroxyls start to react. Therefore, the degree of nitration can be controlled by the length of time the reaction is allowed to continue. Figures 1.34 and 1.35 show the nitration of the single side-chain hydroxyl, and all the hydroxyls, respectively. The weight percent of nitrogen in the completely nitrated cellulose polymer is 14.15%. This is difficult to achieve economically, and other than special laboratory lots, nitrocellulose always has a lower nitrogen content. High-grade nitrocellulose or "guncotton" has a nitrogen weight percent of 13.4. Pyrocellulose is 12.6% nitrogen. Nitrocellulose of much lower percentages of nitrogen (less than 12.0) are used in other plastic products; however, they are not generally suitable for propellant use. Guncotton and pyrocellulose are usually mixed to

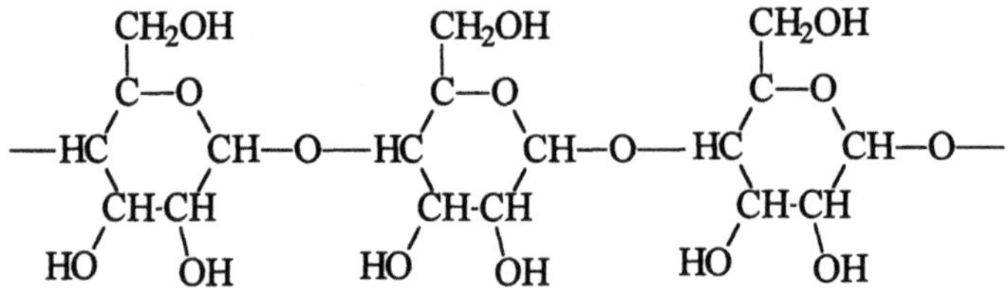

```
     CH2OH                CH2OH                CH2OH
       |                    |                    |
       C—O                  C—O                  C—O
      /   \                /   \                /   \
 —HC       CH—O—HC          CH—O—HC              CH—O—
      \   /                \   /                \   /
      CH-CH                CH-CH                CH-CH
     /     \              /     \              /     \
   HO       OH          HO       OH          HO       OH
```

Figure 1.33. Three monomer links in a nitrocellulose polymer chain.

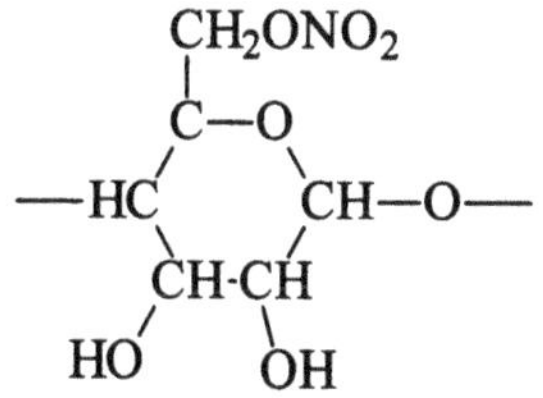

```
        CH2ONO2
        |
        C—O
       /   \
  —HC       CH—O—
       \   /
       CH-CH
      /     \
    HO       OH
```

Figure 1.34. Nitration of the single hydroxyl on the side chain.

yield "military blends" of from 13.15 to 13.25% nitrogen. See Table 1.8 for the weight percent of nitrogen in each of three possible stages of nitration.

As with most nitrate esters, nitrocellulose slowly decomposes at elevated temperatures. During the decomposition, nitric oxide (NO) and nitrogen dioxide (NO_2) are given off. This then reacts with other parts of the chain and accelerates further decomposition. To prevent the secondary reaction (NO_2 with the other parts of the chain), stabilizers are added. The stabilizers react with the NO_2 to form nitrates and nitrate esters and thus prevent further degradation. Several of these stabilizers are in common use. They include: diphenylamine, nitrodiphenylamine, ethyl centralite, ethylaniline, carbazole, nerolin, mineral jelly, and akardit I, II, and III.

These stabilizers are generally present in quantities of less than 2%. Other additives, such as graphite (for static charge suppression and die lubrication) and plasticizers, are also present. During the processing, the nitrocellulose and various additives are dissolved in a solvent, usually acetone. This thins the material sufficiently so that it can be extruded through dies similar to spaghetti extruders. As the strings of propellant exit the die, they are cut off in short lengths, thus yielding cylindrical pellets. Other processes produce sheets, discs, perforated pellets, and spheres. These individual pieces are called "grains." After forming, most of the solvent evaporates from the grains; it is then recovered, condensed, and recycled. Some examples of single-base propellants are shown in Table 1.9.

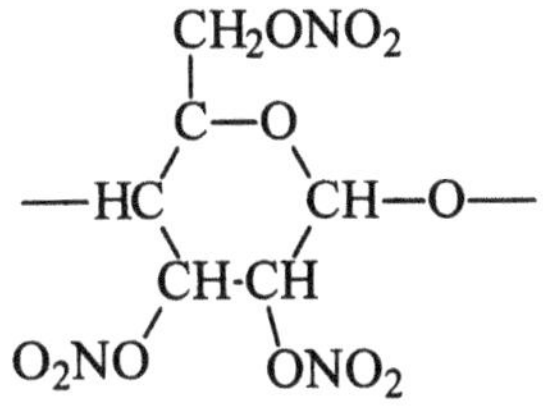

```
        CH2ONO2
        |
        C—O
       /   \
  —HC       CH—O—
       \   /
       CH-CH
      /     \
  O2NO       ONO2
```

Figure 1.35. Complete nitration of all the hydroxyl groups.

Table 1.8 Nitrogen in Single-Base Propellants

Number of —OH Groups Nitrated	Formula	Wt % Nitrogen
1	$C_6H_9NO_7$	6.76
2	$C_6H_8N_2O_9$	11.11
3	$C_6H_7N_3O_{11}$	14.15

1.4.2 Double-Base Propellants

As stated previously, single-base propellants are basically nitrocellulose. Double-base propellants are also made of nitrocellulose but are gelatinized, or plasticized, with another propellant or explosive material that is a liquid. In most cases, the second component in double-base propellants is nitroglycerine. As an alternate to nitroglycerine, however, a large array of short-chain nitrate esters, which were mentioned among the aliphatic explosives, may be used. Dinitrotoluene is also used as the second component. The use of the ''second component'' allows the propellant to be worked and extruded (at elevated temperatures) without having to use a solvent. The second component is intimately bonded to the nitrocellulose and does not separate.

In addition to its function as a plasticizer, the second component is also used to adjust the oxygen balance, and hence affects the energy output and reaction temperature. Nitrocellulose has a negative oxygen balance for CO_2, but is close to being balanced to CO. The positive oxygen balance of the second components, such as nitroglycerine, shifts the overall oxygen balance more toward the positive. The negative oxygen balance second components, such as dinitrotoluene, shifts it even more negative, tending to lower the reaction or burning tempera-

Table 1.9 Properties of Some Single-Base Propellants

	Propellant Types		
Property	M10	M12	Navy Pyro (Pyrocellulose)
Nitrocellulose	98.0	97.7	99.5
% nitration	13.15	13.15	12.95
Oxidizers:			
potassium sulfate	1.0	0.75	
Stabilizers:			
Diphenylamine	1.0		0.5
Lubricants, antistats:			
Graphite	0.1	0.8	
Tin		0.75	

Table 1.10 Properties of Some Double-Base Propellants

Property	Propellant Types						
	M1	M2	M5	M6	M7	M8	M9
Nitrocellulose	85.0	77.45	81.95	87.0	54.6	52.15	57.75
% nitration	13.5	13.25	13.25	13.15	13.15	13.25	13.25
Nitroglycerine		19.5	15.0		35.5	43.0	40.0
Dinitrotoluene	10.0			10.0			
Oxidizers:							
Barium nitrate		1.4	1.4				
Potassium nitrate		0.75	0.75			1.25	1.5
Potassium perchlorate					7.8		
Plasticizers:							
Dibutylphthalate	5.0			3.0			
Diethylphthalate						3.0	
Stabilizers:							
Diphenylamine	1.0			1.0			0.75
Ethyl centalite		0.6	0.6		0.9	0.6	
Lubricants, antistats:							
Graphite		0.3	0.3				
Carbon black					1.2		

ture. The oxygen balance is also adjusted by addition of inorganic oxidizers such as nitrates, perchlorates, or sulfates. An oxygen balance close to CO is important for gun propellants because the resultant muzzle velocity at that condition is at its maximum. The weight percentages of ingredients for some double-base propellants are shown in Table 1.10. The remainder of weight percentages of each propellant is composed of additives indicated by x's in the table.

1.4.3 Triple-Base Propellants

These, like the single- and double-base propellants, are based upon nitrocellulose. Like the double-base ones, they contain a reactive plasticizer (nitroglycerine). They also contain a third major reactive component, nitroguanidine. The nitroguanidine is added to adjust the gas output, energy, temperature, and burning rate. Because the nitroguanidine tends to stiffen the blend, additional nonreactive plasticizers are often added. Some typical triple-base propellants are shown in Table 1.11.

1.4.4 Composite Propellants

The types of propellants we have examined so far are all of the self-oxidizing type. That is, the oxidizer is part of the propellant molecule. In this group of propellants, the composites, the fuel, and the oxidizer are separate materials. The

Table 1.11 Properties of Some Triple-Base Propellants

	Propellant Types					
Property	M15	M16(T6)	M17	M30(T36)	M31(T34)	T20
Nitrocellulose	20.0	55.5	22.0	28.0	20.0	20.0
% nitration	13.15	12.6	13.15	12.6	12.6	13.15
Nitroglycerine	19.0	27.5	21.5	22.5	19.0	13.0
Nitroguanidine	54.7		54.7	47.7	54.7	60.0
Dinitrotoluene		10.5				
Oxidizers:						
Potassium sulfate		1.5				
Stabilizers:						
Diphenylamine					1.5	
Ethyl centralite	6.0	4.0	1.5	1.5		2.0
Plasticizers:						
Dibutylphthalate					4.5	5.0
Lubricants, antistats:						
Cryolite	0.3		0.3	0.3	0.3	
Graphite		0.5	0.1	0.1		
Lead carbonate						1.0

nitrocellulose-based propellants are used primarily as gun and small arms propellants, and in some small military rocket motors. The composite propellants are used for rockets and for gas generators other than in guns. The typical composite is a blend of a crystalline oxidizer (see Table 1.12) and an amorphous or plastic fuel (see Table 1.13). The fuel acts both as fuel and binder. It provides the mechanical strength and structural properties to the composite propellant.

The amorphous fuel systems can be mixed and hot-poured or pressed into molds or molded directly into a rocket motor. They have the disadvantage of melting or softening and creeping at elevated temperatures.

The thermosetting fuel systems are mixed as liquid monomers with the oxidizers and curing agents, poured into molds, and then heated and cured. The

Table 1.12 Typical Composite Propellant Oxidizers

Name	Formula
Sodium nitrate	$NaNO_3$
Potassium nitrate	KNO_3
Ammonium nitrate	NH_4NO_3
Ammonium perchlorate	NH_4ClO_4
Potassium perchlorate	$KClO_4$
Lithium perchlorate	$LiClO_4$

Table 1.13 Typical Composite Propellant Fuels

Name	Form
Asphalt	Amorphous
Polyisobutylene	Polymer, amorphous
Polysulfide rubber	Polymer, thermosetting
Polyurethane rubber	Polymer, thermosetting
Polybutadiene-acrylicco-polymer	Polymer, thermosetting
Polyvinyl chloride	Polymer, thermoplastic
Cellulose acetate	Polymer, thermoplastic

advantage is that once cured they do not change their shape, even at very high temperatures. The disadvantage is cost, and later problems in recycling the ingredients at the end of the useful lifetime of the propellant.

The thermoplastic systems do soften and melt, but at much higher temperatures than the amorphous systems. The thermoplastic systems can be either hot-extruded or injection- and compression-molded.

The energy output, and hence the flame temperature of composite propellants, is controlled by both the choice of fuel and oxidizer and by the fuel/oxidizer ratio. The burning rate is not only a function of fuel/oxidizer ratio but also of oxidizer particle size, propellant density, porosity, and emissivity.

Typical values of burning rates are from 0.1 to 1 in./s (at 1000 psi). Like the other propellants, they produce gas on the order of 1 l/g and energy of approximately 1 kcal/g.

The oldest of the composite propellants is black powder. The name is not because of the color of this material. Black powder is the direct English translation of "Schwarzpulver," named after Berthold Schwarz, who experimented with it in the fourteenth century in Germany. Black powder is made of a combination of carbon (wood charcoal), sulfur, and potassium nitrate. Some black powders, designated as "Grade B blasting," use sodium nitrate instead of potassium nitrate. Military grades JAN A, B, and C also use sodium nitrate.

Black powder is produced by mixing the ingredients wet and then wheel-milling them. The heavy wheel mill (wheels weigh as much as 10 tons) mix and crush the powder. The tremendous pressures in the mill cause the sulfur to plasticize and flow, binding the charcoal and nitrate while in this amorphous state. The milled product is then further treated, dried, and screened. The screening produces the various grades, whose bulkburning rates are a function of the particle sizes.

Black powder burning rates are somewhat higher than those of other propellants at low pressure and are lower at high pressure. It produces gas at around 300 cm^3/g and energy at around 0.7 kcal/g. Although the specific energy of black powder is relatively low, it is nonetheless quite dangerous and ignites easily from sparks, especially when it is in the form of small particles. When used as a small-rocket fuel, black powder is mixed with organic glues (usually water soluble), pressed into the motor, and then dried. Although it is the oldest of the propellants, it is still used extensively. In addition to the commercial uses in fireworks, it is the main ingredient in safety fuses. It is used by the military in small rockets, delay trains, mine projectors, and in gun primers and ignitors.

1.5 Pyrotechnics

Pyrotechnics usually consist of a mixture of two ingredients: a fuel and an oxidizer. Generally, the fuels are metals and the oxidizers are either salts or metal oxides. They differ from explosives in that the reaction rates are less than the bulk sound speed in the unreacted material. In most cases the reaction rates are very low. They differ from propellants in that in most cases relatively little gas is produced. Of course, there are exceptions to these generalities, as we will see.

In both explosives and propellants, we were dealing with materials that were relatively nonporous. The binders and plasticizers used effectively filled the pore spaces. In many pyrotechnics no binders or plasticizers are used. The explosives and propellants have burning rates (or detonation rates) that depend primarily on density, temperature, and pressure. Because the pyrotechnics are porous, the convective heat transfer due to hot gas permeation into the reactant material mixture ahead of the reaction zone becomes important. The burning rates of pyrotechnics, therefore, are not only affected by pressure, density, and temperature, but also by porosity, particle sizes, purity, homogeneity (degree of mixing), and stoichiometry (fuel/oxidizer ratio).

Many substances have been used in the past in pyrotechnic mixtures but are now avoided since they lead to environmental and health problems. These include beryllium, cadmium, mercury, chromates, lead compounds, etc.

Pyrotechnics are divided into five major production categories: sound, light, heat, smoke, and delay.

1.5.1 Sound Producers

Sound-producing pyrotechnics are primarily used in the fireworks industry and in military simulators. There are two types of sound producers: a composition that produces a loud, short-duration sound, like a bomb or grenade, and a composition that produces a shrill whistle of long duration. The short loud bang producers are sometimes made of black powder that is heavily confined in a cardboard tube. More typically, however, sound is produced by photoflash mixes. These consist of various blends containing aluminum, magnesium, potassium perchlorate, and other additives. Some of the blends are so fast that they require little or no confinement in order to explode. Typical sound producers are firecrackers, whistles, salutes, and military grenade and ground-burst simulators.

1.5.2 Light Producers

Basically, there are two subdivisions of light-producing pyrotechnics: flash powders and flares. The difference in performance is in the bulk burning rates. Flash powders burn very rapidly, some almost bordering on detonation velocities. The flash powders are generally loaded as loose or lightly pressed mixed dry powder. They are sometimes initiated by an ignitor and in some applications by a detonator. They are used as time markers in many military and U.S. Department of Energy systems experiments, and also as simulators for a detonation. In military tactical use, they are employed as lighting for night reconnaissance photography.

The majority of photoflash mixes use magnesium and/or aluminum as the fuel and barium nitrate and/or potassium perchlorate as the oxidizers. The typical reactions of the metal/salt mixtures can be exemplified as:

$$8Al + 3KClO_4 \rightarrow 4Al_2O_3 + 3KCl$$

and

$$4Al + 3BaNO_3 \rightarrow 2Al_2O_3 + 3BaO + 1.5N_2$$

Reaction rates for photoflash powders have been reported to be as high as 1 km/s and depend upon, as previously stated, chemical mix, particle size, and pressing density.

Flares, on the other hand, burn very slowly. They are used to produce sustained lighting for a variety of purposes, including signaling and battlefield illumination. They are also metal/salt mixtures, but with a large amount of binder. Typical fuels are magnesium (sometimes mixed with aluminum), manganese, and silicon. Typical oxidizers include the nitrates of barium, sodium, potassium, and strontium. Binders used are castor and linseed oils and paraffin waxes. Another interesting flare mix is magnesium/Teflon. The chlorine and fluorine from the Teflon are the oxidizers in this mix. Magnesium/Teflon flares burn several hundred degrees (°C) hotter than metal/salt flares and radiate very strongly in the infrared spectrum. Burning rates are in the range of 0.25 to 2 in./s.

One of the more common flares used by the private sector is the warning flare that signals oncoming highway traffic during an accident or breakdown of a vehicle. These devices utilize strontium nitrate, potassium perchlorate sulfur, wax or grease, and hardwood sawdust.

Another light producer is a tracer mix. This is very similar to flare compositions, except it may also have additions of other metal salts to enhance or vary the flame color.

1.5.3 Heat Producers

Heat-producing pyrotechnics are used for ignition mixes, thermites, sparks, and incendiaries and heat pellets for thermally activated batteries. The first element, ignition mixes, are generally either metal/salt or metal/metal oxide mixes with no binders. The metal/salt mixtures are very sensitive to impact, flame (or a concentrated heat source such as a glowing hot wire), and sparks, whereas the metal/metal oxide mixtures are not sensitive to impact, flame, or spark. Both types of mixtures are used as the "first fire" or the ignition element in a pyrotechnic train. Typical fuels are aluminum, zirconium, titanium, and titanium hydrides, and magnesium, boron, and in former times, beryllium. Typical oxidizers include the various nitrates already mentioned with the light producers, as well as calcium chromate, lead nitrate, iron oxides, copper oxide, and the perchlorates of sodium, potassium, and ammonia. The calcium chromate and lead nitrate are now considered to cause environmental problems and are falling out of use. A typical reaction of the metal/metal oxide type is the reduction/oxidation of aluminum mixed with copper oxide.

$$2Al + 3CuO \rightarrow Al_2O_3 + 3Cu$$

This reaction, where an oxide of one metal is reduced and the other metal is then oxidized, is the same reaction as in the thermites. These materials, generally a mixture of aluminum and iron oxide, produce molten iron as one of the products.

$$2Al + Fe_2O_3 \rightarrow Al_2O_3 + 2Fe$$

The molten iron is then the heat transfer medium that is used to perform the specific task of the thermite device, usually welding. Other fuels include nickel, and other oxidizers include Fe_3O_4 and Cu_2O.

Spark mixes are generally any of several pyrotechnic blends with large particle size additions of zirconium. When burning, the zirconium particles jump out and continue to burn in air.

Another spark or hot particle producer is Pyrofuze®, which is a metal/metal reaction. In this instance, the metals do not burn, but one dissolves in the melt of the other, generating a large exothermic heat of solution. This process is also called an exothermic alloying reaction. Pyrofuze®-type wires are made of an aluminum core swaged into a tube of palladium through wire-forming dies.

Another type of heat producer is a metal/salt-type mixture used in some detonators as a first fire mix. A common mix for this type is boron/calcium chromate.

Incendiaries are of two types. The first employs phosphorus that is dispersed and then burns with air. The second type is called napalm, which is a mixture of gasoline, thickeners, and gums. The napalm does not ignite merely from contact with air as does the phosphorous, but requires an ignitor.

1.5.4 Smoke Producers

These pyrotechnics are also subdivided into organic and inorganic types. Both forms of smokes are used as markers, signals, and, tactically, as obscurants. They can be made in a variety of colors and optical densities. The earliest organic smokes were merely droplets of kerosene that condensed from vapors produced

Table 1.14 Typical Smokes

Smoke	Fuels and Inerts[a]	Oxidizers
White	Zinc	Hexachlorethane
White	Phosphorous acid	Phosphorous pentoxide
White	White phosphorous	Air
Black	Sulfur and pitch	Potassium nitrate
Black	Pitch (bituminous)	Air
Gray	Zinc dust	Hexachlorethane or naphthalene
Yellow	Auramine and lactose	Potassium chlorate
Red	Rhodamine red, gum arabic, and antimony sulfide	Potassium perchlorate
Red	Red dye, sulfur, and baking soda	Potassium chlorate
Green	Auramine, indigo, and lactose	Potassium chlorate
Blue	Indigo and lactose	Potassium chlorate

[a]Some of the inerts listed react slightly at high temperatures.

from boiling the kerosine. This smoke has the obvious drawback of being extremely flammable. The other organic smokes are quite different. They are aerosol droplets of condensed organic dyes that are vaporized out of a relatively low-temperature smoke mix. Irritant agents such as tear gas are also formed this way. The flame temperature of the burning mix is sufficiently low that the organic dye material does not thermally decompose. The pyrotechnic mixture frequently used to produce this low-temperature burn consists of potassium chlorate and powdered sugar diluted with bicarbonates to cool the reaction. Burning rates of these mixes are low, usually in the range of 0.02 to 0.1 in./s.

The inorganic smokes are from reactions that produce zinc chloride as the major obscurant. These mixes, called HC smokes, are a blend of aluminum powder, zinc oxide, hexachlorethane (HC), and sometimes a pinch or two of magnesium and ammonium perchlorate. At low percentages of aluminum, the major smoke products are Al_2O_3 and $ZnCl_2$ with CO. At higher aluminum concentrations the products form a grey, instead of white, smoke that consists of Al_2O_3, $ZnCl_2$, and carbon.

Another inorganic smoke is formed by the burning of white phosphorus; thus it is called a "WP" smoke. It is used in some hand grenades such as the M15. Several varieties of colored smokes are listed in Table 1.14

1.5.5 Delay Producers

Delay producers are pyrotechnic devices that use the controlled burning rate of a mix to create a predetermined time delay between the ignition event and output. Pressed delay mixes utilize both metal/metal oxide and metal/salt formulations. Cord-type delays, such as safety fuse, quarry cord, and ignitor cord, use mixtures of black powder and a diluent. Of the pressed delays, some produce gas, and in

Table 1.15 Typical Pressed Powder Delays

Fuel	Oxidizer	Burning Rate Range (in./s)
Silicon	Red lead plus filler	0.5–2
Boron	Barium chromate	0.5–2
Zirconium nickel alloy	Barium chromate plus potassium perchlorate plus filler	0.1–1
Manganese	Barium and lead chromate	0.1–2
Tungsten	Barium chromate plus potassium perchlorate plus filler	0.2–1
Chromium	Barium chromate plus potassium perchlorate	0.1–1
Molybdenum	Barium chromate plus potassium perchlorate	0.05–100
Selenium plus tellurium	Barium peroxide	0.1–0.5

order to maintain constant burn rates, these delay columns must be vented to relieve any buildup of internal pressure. The delay mixes that form no gaseous products can be sealed into the delay device. Sealed delays are called "obturated"; unsealed, or vented delays, are called "nonobturated." If delay columns are not intimately bound or sealed to the walls of the column, gas leakage can occur along this interface, causing "flash-throughs." Some typical pressed powder delay mixes are shown in Table 1.15.

1.6 Related Reading

The following reading materials are recommended for a wide perspective of the history, use, and misuse of explosives, as well as a more in-depth study of the chemistry. These materials also provide a broad database for explosives, propellants, and pyrotechnics.

Books

1. Ellern, H., *Military and Civilian Pyrotechnics*, Chem. Pub. Co. (1968).
2. Meyer, R. *Explosives*, Verlag Chemie Press (1977).
3. Conkling, J. A., *Chemistry of Pyrotechnics*, Marcel Dekker Press (1985).
4. Lancaster, M., Shimizu, T., et al., *Fireworks—Principles and Practice*, Chem Pub. Co. (1972).
5. McLain, J., *Pyrotechnics*, Franklin Inst. Press (1980).
6. *Explosives and Rock Blasting*, Atlas Powder Co. (1987).
7. Davis, T., *The Chemistry of Powder and Explosives*, Angriff Press (1943).
8. Cook, M. A., *The Science of High Explosives*, Reinehold Pub. Co. (1958).
9. Fordham, S., *High Explosives and Propellants*, Pergamon Press (1966).

10. Urbanski, T., *Chemistry and Technology of Explosives* (4 volumes), Pergamon Press (1964).

11. Dutton, W., *One Thousand Years of Explosives*, Holt, Rinehart & Winston Pubs. (1960).

12. *Dictionnaire de Pyrotechnie*, Groupe de Travail de Pyrotechnie, France (1993).

13. Liebenberg, D. H., Armstrong, R. W., and Gilman, J. J., *Structure and Properties of Energetic Materials*, Materials Research Society, Pittsburgh, 1992.

Handbooks and manuals

1. Dobratz, B. M. (Ed.), *Lawrence Livermore National Laboratory Explosives Handbook*, UCRL-52997 (Revised Jan. 1985).

2. Brauer, K., *Handbook of Pyrotechnics*, Chem. Pub. Co. (1974).

3. *Blaster's Handbook*, E. I. DuPont de Nemours & Co. (1980).

4. *Properties of Explosives of Military Interest*, U.S. Army Materiel Command AMP706-177, (January 1971).

5. *Military Pyrotechnic Series: Part I*, U.S. Army Materiel Command, AMCP 706-185 (April 1967).

6. *Solid Propellants: Part I*, U.S. Army Materiel Command, AMCP 706-175 (September 1964).

7. *Properties of Materials Uses in Pyrotechnic Compositions*, U.S. Army Materiel Command, AMCP 706-187, (1967).

8. *Encyclopedia of Explosives and Related Items*, PATR-2700, Picatinny Arsenal.

Journals

1. *Journal of Propellants, Explosives, Pyrotechnics* (official journal of the International Pyrotechnics Society), VCH Publ., Wienheim, FRG.

2. *Journal of Energetic Materials*, Dowden, Brodman & Devine Publ., Wharton, NJ.

3 *Journal of the Society of Explosives Engineers.*

Conference proceedings

1. *Symposium (International) on Detonation*, held every 4 years starting in 1951, sponsored mostly by the U.S. Navy.

2. *Conference on Explosives and Blasting Techniques*, sponsored by the Society of Explosives Engineers, held annually since 1975.

3. *Internation Pyrotechnic Seminar*, Sponsored by the International Pyrotechnics Society, held every 2 years starting in 1968 until 1984 and annually since then.

4. *Symposium on Explosives and Pyrotechnics* (formerly called *Symp. on Electroexplosive Devices*), sponsored by the Franklin Institute.

CHAPTER

2

Mechanics of Burning

In this chapter we will examine the dynamics of the burning of a propellant. We will start with a surface burning model that will lead us to the effects of pressure on the normal burning rate. Then we will examine the effects caused or controlled by the shape and size of the propellant, leading us to the calculation of gas production rates. Then, by using the ideal gas equation as the equation of state of the gases, we will see how to predict pressures and temperatures in a fixed closed volume in which a propellant is burning. Finally, we will see how all of the above is coupled with the equations of motion in order to predict how a burning propellant causes a piston to move, which introduces us to the field of interior ballistics.

2.1 Burning Model

Propellants are, for the most part, polymeric or plastic materials. In their use, they are formed into special shapes of fairly high mechanical properties. To understand how they burn, let us consider a mass of propellant burning on one surface only. The entire process of burning consists of several steps, as shown in Figure 2.1 Each of these steps is in dynamic balance, and we assume that the entire process is occurring at some steady state.

At the face of the propellant, heat is being received from the burning reaction by radiation, causing the surface to melt. The melted propellant, receiving still more heat, then boils or evaporates. During both the melting and vaporization stages, the propellant begins to decompose. The decomposition products react

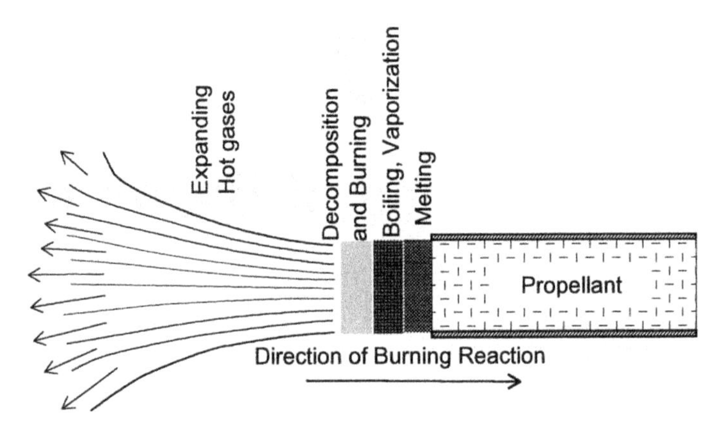

Figure 2.1. Solid propellant burning model.

with each other in the flame region, and here the combustion becomes complete. The product gases from the hot combustion flow away from the surface.

As long as the flame continues to radiate at the same rate, the boiling continues at the same rate, as does the vaporization and decomposition. The entire process then maintains a steady rate, propagating its way into the propellant material at a constant velocity. This velocity with which the reaction proceeds into the propellant is called the normal burning rate.

If we raised the pressure of the gases surrounding the propellant, we would, of course, cause everything in the gas phase to be compressed. This means the flame region would be denser, and hence, would give off more heat per unit volume. The flame also would be pressed closer to the propellant surface. These two factors combined would increase the radiation heat transfer to the surface causing more heat to flow into the surface of the propellant, which, in turn, would force the propellant to melt and vaporize faster. Thus we see, qualitatively, that by increasing the pressure we increase the burning rate. Quantitatively, this process can be expressed in a simple form, the normal burning rate equation,

$$R = aP^n \tag{2.1}$$

where R is the normal burning rate (usually expressed in in./s), at the burning rate coefficient (in./s/psi^n), P the pressure (psi), and n the burning rate exponent (dimensionless). This expression has been shown to be very accurate over most operational ranges of pressure for all the common types of propellants. Table 2.1 gives the normal burning rate coefficients and exponents for a number of common propellants.

Example 2.1

At one atmosphere black powder burns faster than smokeless powder, but at high pressure it is the opposite; smokeless burns faster than black. What are the normal burning rates of black powder and Bullseye (a very common reloading powder for pistol cartridges)

Table 2.1 Normal Burning Rate Coefficients and Pressure Exponents

Propellant	a (in./s)/(psi^n)	n (Dimensionless)	Type	Typical Uses
M1	0.00214	0.700	double base	Artillery
M2	0.00243	0.755	double base	Artillery
M5	0.000761	0.895	double base	Artillery
M6	0.00520	0.674	double base	Artillery
M14	0.00520	0.674	double base	Artillery
M15	0.00330	0.693	triple base	Artillery
M17	0.00690	0.630	triple base	Artillery
T20	0.00330	0.693	triple base	Artillery
M30 (T36)	0.00576	0.652	triple base	Artillery
M31 (T34)	0.00356	0.650	triple base	Artillery
M10	0.00400	0.695	single base	Recoilless rifle
M16	0.00577	0.580	triple base	Recoilless rifle
T18	0.00210	0.780	double base	Recoilless rifle
T25	0.00210	0.780	double base	Recoilless rifle
M26 (T28)	0.00830	0.870	double base	Recoilless rifle
M7 (T4)	0.00247	0.810	double base	Mortar
M8	0.0018	0.830	double base	Mortar
M9	0.00185	0.850	double base	Mortar
IMR	0.00400	0.695	double base	Small arms
M12	0.00400	0.695	single base	Small arms
Bullseye	0.000775	0.920	double base	Small arms
Red Dot	0.00350	0.720	double base	Small arms
Navy Pyro	0.00400	0.695	single base	Small arms
Black powder	0.436	0.165	composite	Various

at one atmosphere (14.7 psi) and at 35,000 psi (a typical small arm breach pressure)? At what pressure do they both have the same normal burning rate? Solution: Using Eq. (2.1), $R = aP^n$, and using the data for a and n for the two propellants from Table 2.1, we get:

$$\text{at 14.7 psi: } R(\text{black powder}) = 0.436(14.7)^{0.165} = 0.68 \text{ in./s},$$

$$R(\text{Bullseye}) = 0.000775(14.7)^{0.92} = 0.0092 \text{ in./s},$$

$$\text{and at 35,000 psi: } R(\text{black powder}) = 0.436(35{,}000)^{0.165} = 2.45 \text{ in./s},$$

$$R(\text{Bullseye}) = 0.000775(35{,}000)^{0.92} = 11.7 \text{ in./s}.$$

In order to find the pressure at which both burn at the same rate, we can equate the two as follows:

$$R = a_1P^{n1} = a_2P^{n2},$$

$$\text{then } (a_1/a_2) = (P^{n2}/P^{n1}) = P^{(n2-n1)},$$

$$\text{and } P = (a_1/a_2)^{1/(n2-n1)} = (0.436/0.000775)^{1/(0.92-0.165)} = 4400 \text{ psi}.$$

Notice that all the propellants listed in Table 2.1 are gun propellants. Rocket motors are usually designed to burn at some given constant chamber pressure, normally at 1000 psi. Therefore, burning rates of rocket propellants are usually

Table 2.2 Normal Burning Rates at 1000 PSI for Rocket Propellants

Propellant	Burning Rate (in./s)	Type
Su	0.370	Double-base
Sc	0.370	Double-base
X-13	0.200	Double-base
T16	0.280	Double-base
T37	0.165	Double-base
OGK	0.270	Double-base
T19	0.400	Double-base
010	0.275	Double-base
N5	0.460	Double-base
X-8	0.665	Double-base
JPN	0.610	Double-base
T14	0.180	NH_4ClO_4 + binder composites
T17	0.290	NH_4ClO_4 + binder composites
T24	0.220	NH_4ClO_4 + binder composites
T13E1	0.350	NH_4ClO_4 + binder composites
M20	0.520	NH_4ClO_4 + binder composites
T22	0.410	NH_4ClO_4 + binder composites
GCR-201C	0.380	NH_4ClO_4 + binder composites
T27	0.500	NH_4ClO_4 + binder composites
AN-534J	0.370	NH_4ClO_4 + binder composites
ARCITE 368	0.370	NH_4ClO_4 + binder composites
ANP2541CD	0.430	NH_4ClO_4 + binder composites

only reported at 1000 psi, and the constants for the rate equation are often not available. Normal burning rates at a constant pressure of 1000 psi for a number of common rocket propellants are shown in Table 2.2.

The above burning rate data for the gun and rocket propellants were given at ambient temperature 70°F (21°C). That is the temperature of the grain before the start of burning. As mentioned before, nitrocellulose propellants are polymers or plastics; the composite fuels are also plastics and plastic-type materials. All of these are very good thermal insulators. Therefore, when burning occurs, very little heat is transferred beyond the immediate surface into the grain, and the grain maintains its original ambient temperature for the entire burning process. Figure 2.2 shows the steady-state temperature gradients typical in most propellants during burning.

If the propellant grain started at a higher ambient temperature, less heat would be required to bring the surface temperature to the melting point. Therefore, the burning process would go on somewhat faster. The effect of initial temperature on normal burning rate is relatively small, but can be compensated for by the equation for the burning rate temperature correction

$$R = R_o e^{\sigma(T-T_o)} \tag{2.2}$$

where R is the normal burning rate at the new temperature T, R_o the rate at T_o (normally 70°F), e the Euler-Hermite number 2.718, and σ the temperature coef-

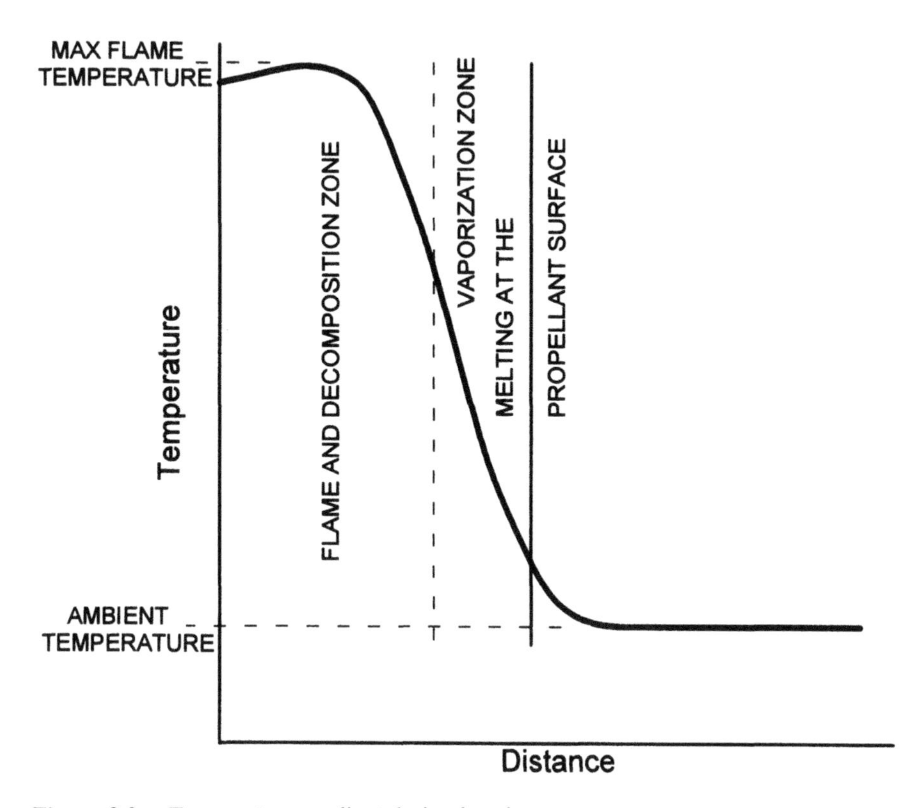

Figure 2.2. Temperature gradient during burning.

ficient (note that for this equation T must be in °F). The value of σ is small (~0.002) for most of the common propellants. Therefore, $e^{\sigma(T-T_o)}$ is very close to one over a fairly broad temperature range. Typically, the R is increased by 10% for a 50°F rise in ambient temperature.

Example 2.2

If we were designing an artillery piece for the Army, we would have to be able to design to the extremes of the anticipated environmental conditions. Suppose we are going to use M30 powder. What are the normal burning rates at nominal temperature (70°F) and at the MIL-SPEC extremes −60°F and +160°F? (The anticipated breach pressure at which we are interested in the rates is 50,000 psi.)

Solution At nominal temperature, we use the normal burning Eq. (2.1) with data for M30 from Table 2.1.

$$R = aP^n = 0.00576(50{,}000)^{0.652} = 6.67 \text{ in./s.}$$

To find the rate at the temperature extremes, we use the nominal rate just calculated and correct it using Eq. (2.2).

$$R = R_o e^{\sigma(T - T_o)},$$

for $T = -60°F$, $R = 6.67\ e^{0.002(-60-70)} = 5.14$ in./s,

and at $T = 160$, $R = 6.67\ e^{0.002(160-170)} = 7.99$ in./s.

2.2 Geometry: Shapes of Grains

The normal burning rate is only part of the story of how a propellant burns. We saw that burning occurs at the surface of a propellant; therefore, the shape and size of that surface is important. If we had a cube subdivided into three equal units along one edge, as shown in Figure 2.3, the total surface area would be 3 × 3 × 6 square units of surface (54 units^2), and the volume would be 3 × 3 × 3 cubic units (27 units^3).

If we divided the cube along the dotted lines, we would have 27 equal cubes, the volume of each being 1 × 1 × 1 or 1 cubic unit (Figure 2.4). The total volume is still 1 × 27 or 27 cubic units. The surface area of each cube is 1 × 1 × 6 or 6 square units. Therefore, the surface area of 27 cubes is 6 × 27 or 162 square units. We have increased the surface area by three times: 162 square units as compared to 54 square units originally.

If this cube were made of propellant, it is easy to see that the group of 27

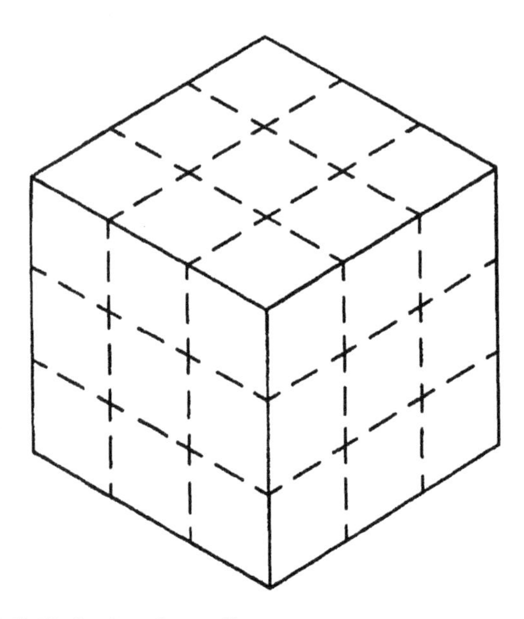

Figure 2.3. Subdivided cube of propellant.

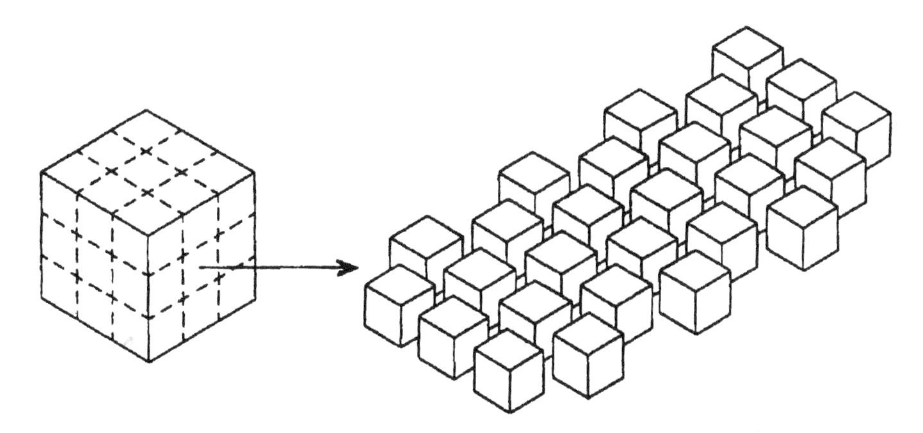

Figure 2.4. Smaller cubes from subdivision of the large cube of propellant.

small cubes would be forming gases from surface burning at three times the rate of the larger cube. Also, the small cubes, since their dimensions are one-third smaller, would burn up in one-third the time. It is obvious that propellant grain size is quite important in determining the rate of gas generation. The rate at which the gas is produced determines the shape of the pressure pulse, or the area under the pulse curve. This is important when designing, for example, a simple gas generator or a projectile for an artillery piece.

The shape of the propellant is also important. The amount of propellant consumed per unit time, or the rate at which gas is formed, is a function of the surface area that is burning times the distance into the propellant burned per unit time. The distance burned per unit time is, of course, the normal burning rate. Therefore, the bulk burning rate is, at any instant, the area that is burning times the normal burning rate or

$$B = AR \tag{2.3}$$

where B is the bulk burning rate (in.^{3}/s), A the surface area burning (in.^{2}), and R the normal burning rate (in./s.)

For a "cigarette-burning" grain, that is, a cylindrical grain only allowed to burn on one surface, it is easy to see that B is a constant (R is held constant) as the grain burns (Figure 2.5). The outer surface is inhibited (a term that will be discussed later) so that only the front surface can burn.

For a hollow cylinder or perforated grain, where only the inside surface is allows to burn, B increases with time (Figure 2.6). For a "ball" grain or sphere, B decreases with time, as the grain burns (Figure 2.7). These three modes of burning characterize the particular grain shape. If B increases with time, the grain geometry is said to be progressive. If B remains relatively constant, the

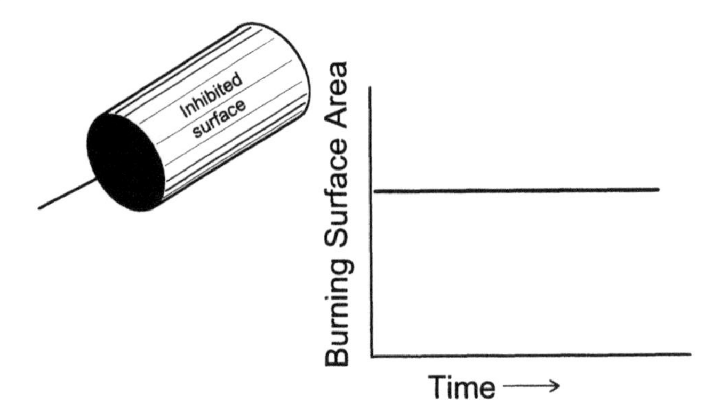

Figure 2.5. Combustion of a cylindrical grain of propellant.

grain geometry is neutral. If B decreases with time, the grain geometry is degressive or regressive.

Figure 2.8 shows some typical geometries of propellant grains that are used in guns and small arms. All gun and mortar propellants burn on all exposed surfaces. In rocket propellants, some surfaces are painted with a nonburning coating, or are bonded to some other nonburning surface. These surfaces, which are not allowed to burn, are called inhibited. Controlled burning of the propellant in rocket motors requires minimizing variation in the propellant burning area and the pressure produced by the burn. Some typical rocket grain geometries are shown in Figure 2.9.

The distance into the grain through which burning takes place until the grain is consumed is one-half of the web. Referring to Figure 2.8, the full web of a

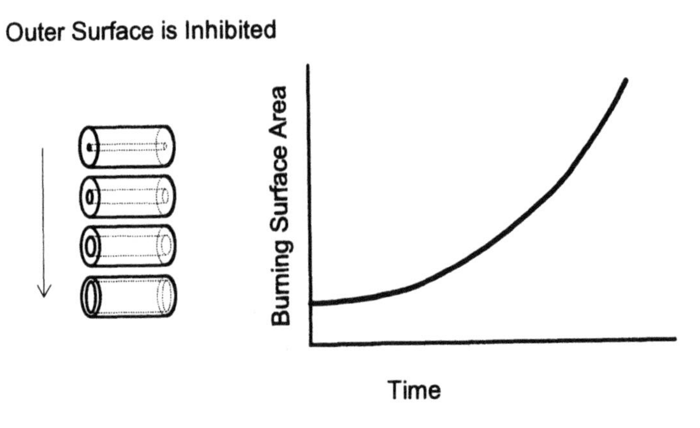

Figure 2.6. Combustion of a perforated grain of propellant.

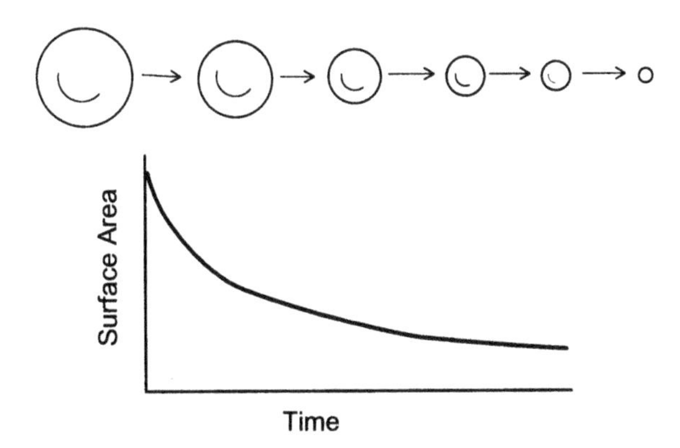

Figure 2.7. Combustion of a spherical grain of propellant.

ball grain is its diameter. The same is true for a cylindrical grain. For both sheets and flakes, the web is equal to the thickness. For single-perforated grains, the web is the wall thickness, or

$$w = (D - H)/2, \tag{2.4}$$

where w is the web (in.), D the grain diameter (in.), and H the perforation diameter (in.). A little simple geometry yields the web for a seven-perforated grain as

$$w = (D - 3H)/4 \tag{2.5}$$

GRAIN	NAME	TYPE	USE
	Ball	Degressive	Small arms
	Flake	Degressive	Small arms
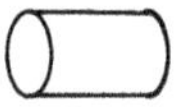	Cylinder	Degressive	Small arms
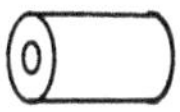	Single perf	Neutral	Rifles and small arms
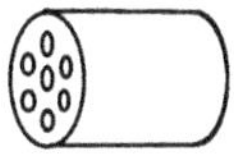	Multi perf	Progressive	Large bore guns
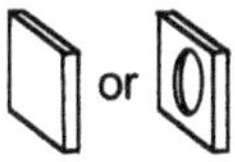 or	Sheet	Neutral	Mortars

Figure 2.8. Gun grain geometries.

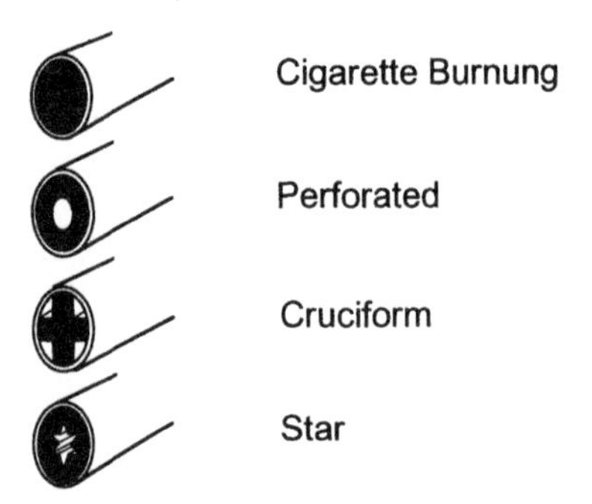

Figure 2.9. Rocket propellant grains.

At constant pressure, the total burning time of a grain is equal to one-half the web times the normal burning rate (at that pressure).

In general, for guns, the web of the propellant (and therefore the size of the grain) increases proportionally with the bore of the gun or the diameter of the projectile. Figure 2.10 shows the propellant web for a number of different seven perf grains used in U.S. Navy guns.

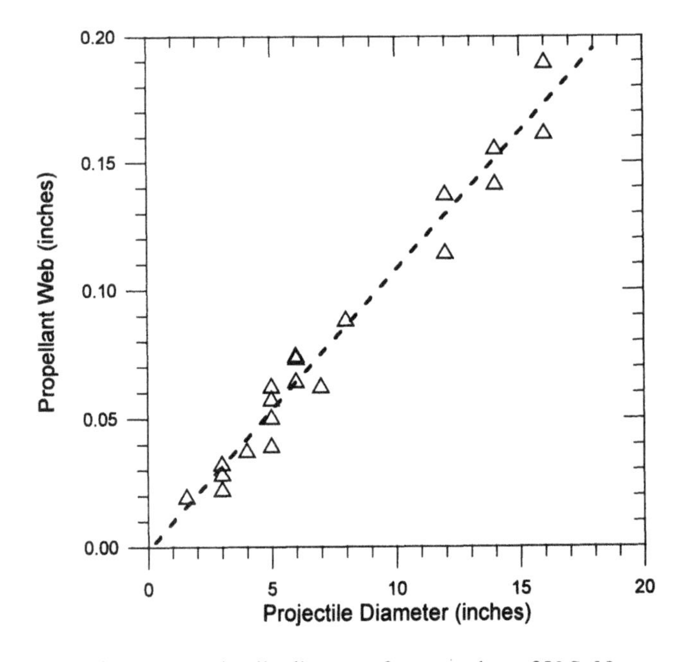

Figure 2.10. Web versus projectile diameter for a number of U.S. Navy guns.

Example 2.3

What is the total burning time for a multiperf (7 holes) grain of M17 propellant that has a 0.5-inch diameter and has 0.032-inch-diameter perforation, and will be burned at a constant pressure of 43,000 psi?

Solution

a. The normal burning rate is [from Eq. (2.1) and Table 2.1]

$$R = aP^n = 0.0069(43{,}000)^{0.63} = 5.73 \text{ in./s.}$$

b. The web for a seven-perf grain is found from Eq. (2.5),

$$w = (D - 3H)/4 = (0.5 - 3 \times 0.032)/4 = 0.101 \text{ in.}$$

c. The total burn time is now found from the normal burning rate in (a) and one-half of the web found in (b).

$$t = (w/2)/R = (0.101/2)/5.73 = 0.00881 \text{ s.}$$

We defined B, the bulk burning rate, as the volume of solid propellant consumed per unit time. We need to know the amount of gas that is formed, because that is what does the work in a propellant system. The amount of product gas formed during burning is, of course, a function of the chemical composition of the propellant. For single-compound, pure-propellant materials, this is fairly straightforward to calculate. For mixed materials, the chemical balance tend to get somewhat involved, and it is easier to obtain the gas output from a reference table. Such tables usually list the gas output in terms of moles of gas per gram of propellant. Table 2.3 lists this quantity for some common gun propellants.

The moles of gas, as shown in Table 2.3, can be converted to volume of gas at standard temperature and pressure (STP, 0°C and 1 atm) by multiplying the volume of gas per mole (22.414 liters) by the number of moles. This assumes that the gases are ideal, which is a fairly good assumption because the gases are low in number of atoms per molecule (CO, H_2O, N_2, etc.) and are at high temperature. Both these factors tend to make the gases more nearly ideal.

Now that the bulk burning rate, the geometry of the grain, and the density and gas output of the propellant are known, the rate of gas production of the system can be calculated.

$$V_{go} = n_p G B_v \rho \tag{2.6}$$

where V_{go} is the amount of gas produced per unit time (at STP), n_p the number of propellant grains, G the moles of gas per gram of propellant, B_v the volume of propellant consumed per unit time per grain of propellant, and ρ the density of the unburned propellant. Combining this with the definition of B_v, we get

$$V_{go} = nG\rho AR \tag{2.7}$$

Table 2.3 Gas Output for Mixed Propellant Formulations

Propellant	G (Moles of Gas per Gram)	Density (lbs/in.^{3})
M1	0.04533	0.0567
M2	0.03900	0.0597
M5	0.03935	0.0596
M6	0.04432	0.0571
M14	0.04338	0.0582
M15	0.04645	0.0600
M17	0.04336	0.0603
T20	0.04794	0.0548
M30 (T36)	0.04308	0.0567
M31 (T34)	0.04619	0.0595
M10	0.04068	0.0602
M16 (T6)	0.04307	0.0570
T18	0.04219	0.0588
T25	0.04133	0.0585
M26 (T28)	0.04157	0.0585
M7 (T4)	0.03543	0.0610
M8	0.03711	0.0581
M9	0.03618	0.0578
IMR	0.04191	0.0602
M12	0.04037	0.0600
M18	0.04457	0.0576
Bullseye	0.03700	0.0590
Red Dot	0.03700	0.0590
Pyro	0.03964	0.0566
Black Powder	0.01250	0.0580

and A, the surface area of the grain at that particular time, is the only factor that ties this to the grain geometry. Since the burning rate (R) is usually reported in inches per second and the density ρ in pounds per cubic inch, an appropriate factor is figured into Eq. (2.7) to yield V_{go} in cubic inches of gas at STP per second. The dimensionally consistent result is

$$V_{go} = 6.21 \times 10^5 nG\rho AR \tag{2.8}$$

where V_{go} is cubic inches of gas at STP (0°C, 1 atm) per second, n the number of grains of propellant, G the moles of gas per gram, ρ the pounds per cubic inch, A the surface area per grain of propellant (in.^{2}), and R is in inches per second. Equation (2.8) yields the rate of gas output at one point in time. When this equation is used, a computer may be required to integrate the rate over the changing geometry and the time period of burning.

It is simpler to calculate the amount of gas evolved from burning a certain amount of propellant. This is

$$V_{go} = WG \tag{2.9}$$

where V_{go} is the total gas evolved, W the total weight of propellant, and G the moles of gas per gram of propellant. To put this in a form with consistent units,

$$V_{go} = 6.21 \times 10^5 WG \qquad (2.10)$$

where V_{go} is in cubic inches of gas at STP, W in pounds, and G in moles per gram.

Example 2.4

How much gas (at STP) is produced by burning 1 ounce of M10 propellant?

Solution

$$V_{go} = 6.21 \times 10^5 WG$$

a. 1 oz $\times$ 1 lb/16 oz = 0.0625 lb.
b. $V_{go} = 6.21 \times 10^5 \times 0.0625 \times 0.4068 = 1579$ in.^{3} (STP)

Now we know how much gas has been produced and/or the rate at which it was produced. This information is in the form of that volume of gas at standard conditions (0°C, 1 atm). Of course, the gas products are not formed at standard conditions. They are at a very high temperature and, if they are confined, are possibly at high pressure.

2.3 Calculating the State of the Gas

To determine the final state of the gas, we must turn to the thermodynamic equations of state. If we assume that the burning reaction has proceeded rapidly and there was no time for any appreciable heat loss to the surroundings, then we can assume that the reaction is adiabatic. For that case the product gases will be the adiabatic flame temperature. In a constant volume process, that is, one in which the gases remain compressed to the original solid propellant density as they are formed, the adiabatic flame temperature is called the isochoric flame temperature, T_v. If the gases are allowed to expand at constant pressure as they are formed and the volume is not contained, then the adiabatic conditions yield the gases at the isobaric flame temperature, T_p. These two values are related to each other by the quantity $\gamma = C_p/C_v$, the ratio of specific heats.

$$T_v = \gamma T_p \qquad (2.11)$$

Table 2.4 shows the flame temperature and γ as well as the heat of reaction for the same group of gun propellants listed earlier in Tables 2.1–2.3.

The ideal gas law related the pressure, volume, quantity, and temperature of a gas,

$$Pv = n\mathcal{R}T \qquad (2.12)$$

Table 2.4 Flame Temperature and γ of Gun Propellants

Propellant	T_v (K)	γ	Δh_{exp}(kcal/g)
M1	2417	1.2593	0.700
M2	3319	1.2238	1.080
M5	3245	1.2238	1.047
M6	2570	1.2538	0.758
M14	2710	1.2496	0.809
M15	2594	1.2557	0.799
M17	3017	1.2402	0.962
T20	2388	1.2591	0.712
M30 (T36)	3040	1.2485	0.974
M31 (T34)	2599	1.2527	0.818
M10	3000	1.2342	0.936
M16 (T6)	2362	1.2540	0.886
T18	2938	1.2421	0.910
T25	3071	1.2373	0.962
M26 (T28)	3081	1.2383	0.955
M7 (T4)	3734	1.2112	1.280
M8	3695	1.2148	1.244
M9	3799	1.2102	1.295
IMR	2835	1.2413	0.868
M12	2996	1.2326	0.933
M18	2577	1.2523	0.772
Bullseye	3780	1.2523	
Red Dot	3208	1.2400	
Pyro	2487	1.2454	1.005
Black Powder	2800	1.1265	0.720

where P is the pressure, v the volume, n the number of moles of gas, $\mathcal{R}$ the universal gas constant, and T the absolute temperature. The value of $\mathcal{R}$ is 0.08205 (liter · atmospheres/gmole · K). To relate the conditions of a gas at one state to the conditions at another state, we use the ideal gas law in the form of the ratio of the two states,

$$(P_1 v_1/P_2 v_2) = (n_1 \mathcal{R} T_1/n_2 \mathcal{R} T_2) \tag{2.13}$$

$\mathcal{R}$ cancels out, and we are left with

$$(P_1 v_1/P_2 v_2) = (n_1 T_1/n_2 T_2) \tag{2.14}$$

This form of the ideal gas law will be most useful in applying what we have derived thus far.

Example 2.5

Let us examine a case where a given amount of propellant is burned in a closed bomb, for example, 1 pound of M1 propellant. After the propellant has burned, the bomb is filled with a hot, high-pressure gas. The bomb volume is 1 cubic foot (1728 cubic inches). What are the final pressure and temperature?

Solution The solution steps are as follows:

1. The volume of gases formed from burning at STP is:

$V_{go} = 6.21 \times 10^5 WG,$

$W = 1$ pound, and

$G = 0.04533$ (from Table 2.3); therefore:

$V_{go} = (6.21 \times 10^5)(1)(0.04533) = 28{,}150$ cubic inches (STP)

2. Since this is a confined, constant-volume burn, the gas temperature is the isochoric flame temperature, which Table 2.4 gives as 2417 K.
3. Therefore, 28,150 in.^{3} of gas at standard conditions (273 K and 1 atm) must be squeezed into 1728 in.^{3}, and at a temperature of 2417 K.

$P_1 = 1$ atm

$v_1 = 28{,}150$ in.^{3}

$T_1 = 273$ K

$v_2 = 1728$ in.^{3}

$T_2 = 2417$ K

$n_1 = n_2$

$P_2 = ?$

Solving Eq. (2.14) for P_2 yields:

$$P_2 = (P_1 v_1 n_2 T_2 / v_2 n_1 T_1)$$

Since $n_1 = n_2$ (we did not lose or gain any gas),

$$P_2 = (P_1 v_1 T_2 / v_2 T_1) = (1 \times 28{,}150 \times 2417)/(1728 \times 273) = 144 \text{ atm}$$

and since 1 atm = 14.7 psi, $P_2 = 144 \times 14.7 = 2{,}117$ psi.

How about that! Now, if the bomb cools to room temperature, say 25°C (298 K), the gas loses its heat. It goes from the state of 144 atm, 2417 K, 1728 in.^{3}, to $P_3 = ?$ at 1728 in.^{3} (it is still in the bomb) and 298 K.

$$P_3 = (P_2 v_2 T_3 / v_3 T_2), \text{ and } v_3 = v_2, \text{ and so}$$

$$P_3 = (P_2 T_3 / T_2) = (144 \times 298/2417) = 17.8 \text{ atm (or 261 psi)}$$

This example is the simplest type of calculation in the interior ballistics field.

2.4 Interior Ballistics

Essentially, interior ballistics refer to the behavior of the burning system inside a gun, rocket motor, gas generator, or closed bomb. In the example we used above, the volume of the system, as well as the amount of gas formed, was a constant.

In a gun, the volume changes once the bullet starts to move down the barrel.

The simple system now starts to get complicated by the fact that the gas is working on the bullet. It is imparting kinetic energy to the bullet, as defined by Newton's second law as $F = ma$. This energy comes from the hot gases that are cooled by the process. The volume is changing because the bullet is moving at a velocity determined by the acceleration. This in turn is dropping the pressure of the gas. As you can see, this gets complex quite fast.

The process cannot be solved by closed-form equations because of the feedback, or interrelation of the quantities of pressure, volume, temperature, and mass of gas, all related to the motion of the piston or bullet. Therefore, iterative, short-time-step, finite-difference methods are used in computer programs to follow the entire process.

In the case of a rocket motor, the volume of the system is changed slowly by the volume of propellant that is being consumed. The mass, or quantity, or gases present in the rocket motor changes because of the discharge of gas through the nozzle to yield thrust.

As can be seen from the preceding discussion, combustion and interior ballistics can become extremely complex when considering the dynamics of a complete system. In all these calculations, however, the system is always described through a combination of the relatively simple relations that follow:

a. The rate of burning as a function of pressure,
b. The rate of gas formation as a function of the geometry of the propellant grain,
c. The change in propellant geometry caused by normal burning as a function of time,
d. The ideal gas equation, or some other appropriate equation of state,
e. The gas flow and mass balance in the system,
f. The changes in volume due to the combustion of solid propellant,
g. The changes in volume due to motion of some part of the system such as a piston or bullet,
h. Forces due to friction, such as between bullet and rifling, and its dependance upon velocity, and
i. Heat transfer and cooling through system boundaries, such as the walls of the gun breach and barrel.

2.5 Related Reading

1. *Design for Projection*, U.S. Army Materiel Command, AMCP 706-247 (July 1964).
2. *Propellant Actuated Devices*, U.S. Army Materiel Command, AMCP 706-270 (Aug. 1963).
3. *Interior Ballistics of Guns*, U.S. Army Materiel Command, AMCP 70-150 (Feb. 1965).
4. *Elements of Aircraft and Missile Propulsion*, U.S. Army Materiel Command, AMCP-285 (July 1963).
5. C. L. Farrar and D. W. Leeming, *Military Ballistics—A Basic Manual*, Brassey's Publishers (Pergamon Press), Oxford (1983).
6. Hayes, T. J., *Elements of Ordnance*, John Wiley & Sons (1938).

CHAPTER 3

Sound, Shock, and Detonation

In this chapter we shall examine pressure waves: sound, shock, and detonation. We shall see how they differ from each other and shall acquire some basic calculation skills in determining or estimating such quantities as wave velocities and shock and detonation pressure. We shall see how density as well as size affects detonation. Then we shall examine six common empirical tests that are used to measure and compare the detonation output characteristics of explosives.

3.1 Material Distortion: Stress and Strain

When a force is applied to a material, the material undergoes a change in shape and/or volume. The force applied upon a surface of the material, force per unit area, is called *stress*. The stress causes the material to distort or change its shape, this distortion is called *strain*. There are three basic types of strain that can occur: simple shear, pure dilatation (volumetric), and uniaxial (in one direction only) stretching. For small stresses, the strain produced in the material is reversible so that, when the stress is removed, the material returns to its original undistorted size and shape. The range of stress for which the strain is completely reversible is called the *elastic* range for that particular material. The stress beyond which a material is no longer elastic and begins to undergo permanent deformation is called the *elastic limit*. The permanent strain caused by stresses above the elastic limit in a material is called *plastic* strain. In this plastic region a solid material can behave like a fluid. Figure 3.1 shows the relationship between stress and specific volume (specific volume is the reciprocal of density) for a typical solid

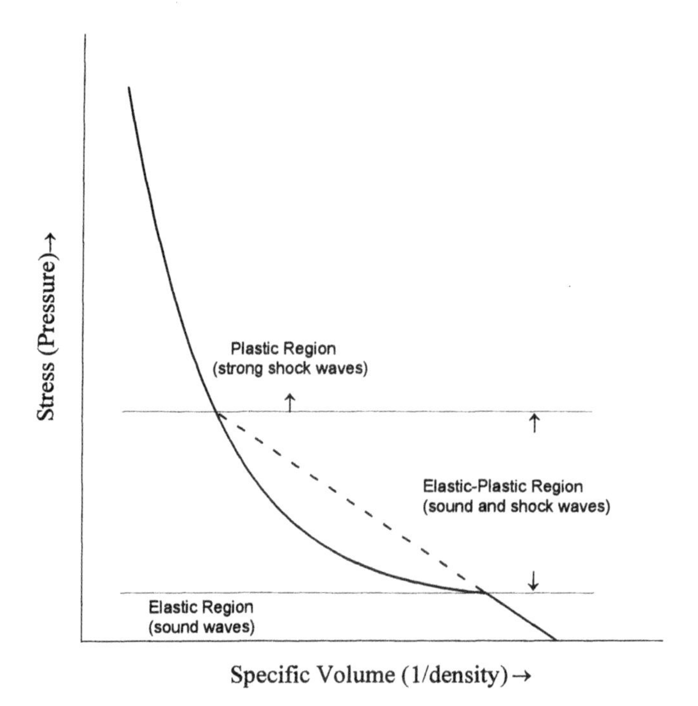

Figure 3.1. Stress-strain relationship for a typical solid material over a very broad range of stress

material. Note that at low stresses, in the elastic region, only sound waves can exist. At very high stresses, in the plastic region, only shock waves can exist. At stresses in the elastic-plastic region, both sound and shock waves can exist at the same time and travel at different speeds.

3.1.1 Elastic Moduli

Within the elastic range the ratio of stress to strain is constant. This relationship is known as *Hooke's Law* and is simply written as: (stress)/(strain) = (constant). Since there are three basic forms of strain, there are three different constants (or moduli) that describe the elastic behavior of a material.

3.1.1.1 Elastic Shear Modulus

Figure 3.2 shows *shear strain*, which is caused by a stress τ applied along a plane.

The elastic shear strain β is the distance of shear x divided by the distance h over which the shear has been applied, or $\beta = x/h$. For purely elastic shear Hooke's Law is expressed as $G = \tau/\beta$ and G is called the *elastic shear modulus*

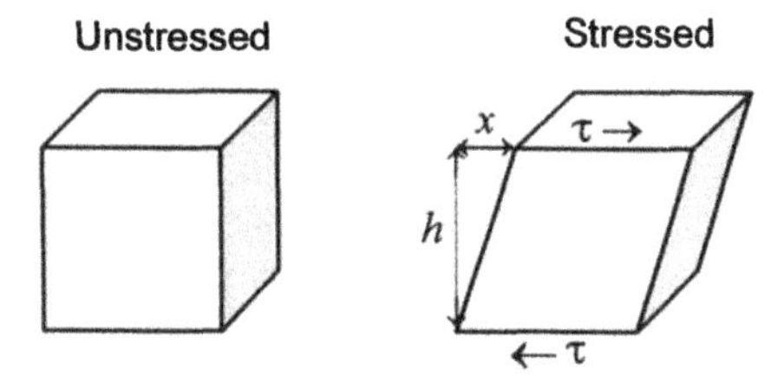

Figure 3.2. Simple shear.

(also called the *rigidity* modulus). It is important to note that pure shear does not change the volume of the subjected material; it only changes the shape.

3.1.1.2 Bulk Modulus

Figure 3.3 shows pure dilatational (volumetric) strain in a material. This would be caused by the application of a uniform stress or pressure σ to all six of the surfaces of the subjected material. In this case the strain is defined as the ratio of the change in volume ΔV to the original volume V, or $\delta = \Delta V/V$. Hooke's Law in this case is stated as $B = \sigma/\delta$, and B is called the *bulk modulus*. It is important to note that pure dilatation does not change the shape of the subjected material; it only changes the volume.

3.1.1.3 Modulus of Elasticity (Young's Modulus)

Figure 3.4 shows tensile strain developed in a material sample subjected to a tensile or pulling stress σ. In this case the strain ε_1 is defined as the ratio of the change in length ΔL of the subjected material to its original length L, or $\varepsilon_1 = \Delta L/L$. Hooke's Law for this case is $E = \sigma/\varepsilon_1$, and E is called the *modulus of elasticity* or *Young's modulus*. Note that the Young's modulus is only for the change in length of the subjected sample. When a sample of material is pulled in tension and elongates, the side or lateral dimensions decrease. The lateral

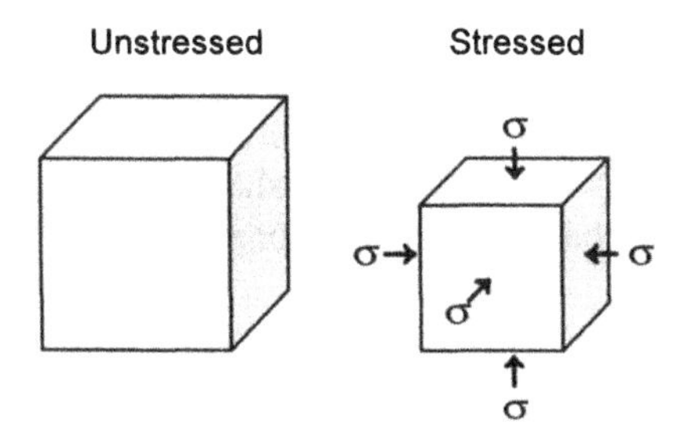

Figure 3.3. Pure dilatation or compression.

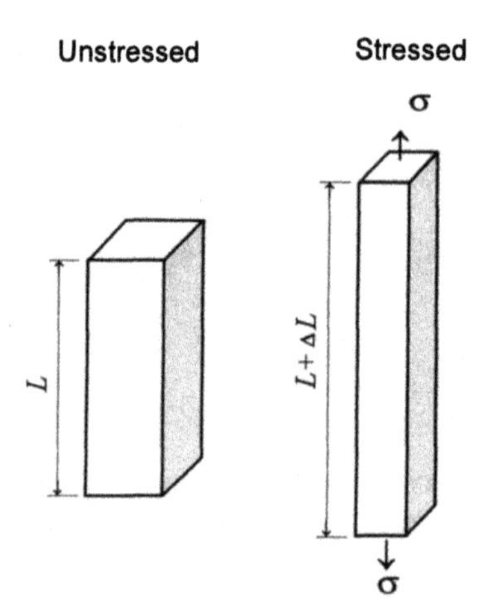

Figure 3.4. Uniaxial tension.

strain is defined in a similar manner as the axial strain, thus yielding ε_x and ε_y, the strains in the two side directions. The ratio of the lateral strain to the axial strain is called *Poisson's ratio* v, where $v = \varepsilon_x/\varepsilon_1 = \varepsilon_y/\varepsilon_1$.

Table 3.1 gives values of the three moduli and Poisson's ratio for several common solid materials.

3.2 Sound Waves

A sound wave is a low-pressure disturbance that is propagated from particle to particle through a material. The pressure disturbance is small, and after the passage of the sound wave, the material returns to its original state, unchanged. The pressures involved have not stressed the material beyond its elastic limit. In solid materials two types of sound waves can be formed by the pressure disturbance depending on how the stress was applied to the material, either as a shear (forming a shear or distortional wave) or a compression (forming a compressive or dilatational wave). Both types of waves can exist in the material at the same time. When a dilatational wave meets a free surface or a density dicontinuity or boundary, both types of wave are reflected. The dilatational wave causes compression in the material as it progresses; the shear wave merely distorts the shape. The two types of waves have different propagation velocities.

3.2.1 Sound Wave Velocity

The dilatational wave is also called a longitudinal wave because the particle motion is along the direction of propagation. The dilatational wave velocity C_L

Table 3.1 Elastic Properties of Various Solids

Material	Density (g/cm^3)	Shear Modulus (n/m^2) $\times 10^{-10}$	Bulk Modulus (n/m^2) $\times 10^{-10}$	Young's Modulus (n/m^2) $\times 10^{-10}$	Poisson's Ratio
Elements					
Aluminum	2.697	2.5	7.77	6.95	0.355
Copper	8.92	4.6	16.17	12.45	0.37
Iron (cast)	7.79	5.99	10.91	15.2	0.27
Lead	11.36	0.54	3.66	1.6	0.43
Magnesium	1.74	1.62	3.64	4.24	0.306
Nickel	8.75	8.0	21.73	21.4	0.336
Alloys					
Brass	8.6	3.8	13.83	10.4	0.374
Stnls. steel	7.91	7.57	16.35	19.6	0.3
Glasses					
Fused silica	2.2	3.12	3.69	7.29	0.17
Pyrex glass	2.32	2.5	3.97	6.2	0.24
Borate crown	2.243	1.81	3.41	4.61	0.274
Polymers					
Lucite®	1.182	0.143	0.66	0.4	0.4
Polyethylene	0.90	0.026	0.31	0.076	0.458
Polystyrene	1.056	0.12	0.42	0.528	0.405

depends upon both the bulk and shear moduli as well as on the density ρ of the solid material

$$C_L = \sqrt{\frac{B + \frac{4}{3} G}{\rho}}$$

The shear wave is also called a transverse wave because the particle motion is perpendicular to the direction of propagation. The shear wave velocity C_s depends only upon the shear modulus and density

$$C_s = \sqrt{\frac{G}{\rho}}$$

Liquids and gases do not exhibit shear properties, and therefore shear waves cannot exist in them. The compressibility K of a liquid is analogous to the bulk modulus of an elastic solid. The velocity of propagation in liquids depends upon the compressibility of the liquid and its density

$$C_L = \sqrt{\frac{K}{\rho}}$$

The compressibility of gases is proportional to the pressure of the gas. The velocity of propagation of a sound wave in a gas depends upon the pressure P, the density ρ, and the ratio of the specific heats γ (where $\gamma = c_p/c_v$, c_p and c_v are the specific heats at constant pressure and constant volume)

$$C_L = \sqrt{\frac{\gamma P}{\rho}}$$

Table 3.2 shows the sound wave velocities for a number of solids (unreacted explosives listed in this table are shown in italics).

Table 3.3 shows the sound wave velocities of various liquids and gases. Unreacted explosives in this table are shown in italics.

3.2.2 Attenuation of Sound Waves

As a sound wave travels farther and farther through a material, the compression and resultant heating it causes represents a loss in energy, referred to as internal friction. Due to internal friction, pressures and densities generally drop over distance traveled to lower and lower levels, close to the ambient, and finally disappear. This process is called attenuation (see Figure 3.5).

Table 3.2 Sound Velocities for Various Solids Including Explosives

Material	Density (g/cm^3)	C_L (km/s)	C_s (km/s)
Aluminum	2.697	6.42	3.04
Copper	8.92	5.01	2.27
Iron (cast)	7.79	5.96	3.24
Lead	11.36	1.96	0.69
Magnesium	1.74	5.77	3.05
Nickel	8.75	6.04	3.00
Brass	8.6	4.70	2.11
Stnls. steel	7.91	5.79	3.10
Fused silica	2.2	5.968	3.764
Pyrex glass	2.32	5.64	3.28
Borate crown glass	2.243	5.10	2.84
Lucite®	1.182	2.68	1.10
Polyethylene	0.90	1.95	0.54
Polystyrene	1.056	2.35	1.12
Baratol (cast)	*2.611*	*2.95*	*1.48*
Comp B (cast)	*1.726*	*3.12*	*1.71*
Cyclotol (cast)	*1.752*	*3.12*	*1.69*
Octol (cast)	*1.80*	*3.14*	*1.66*
PBX 9404 (pressed)	*1.840*	*2.90*	*1.57*
PBX 9502 (pressed)	*1.88*	*2.74*	*1.38*
TATB (pressed)	*1.87*	*2.00*	*1.18*
Tetryl (pressed)	*1.68*	*2.27*	*1.24*
TNT (cast)	*1.624*	*2.48*	*1.34*
TNT (pressed)	*1.632*	*2.58*	*1.35*

Table 3.3 Sound Velocities in Various Liquids and Gases Including Some Explosives

Liquids (at 25°C)	Density (g/cm^3)	C_L (km/s)	Gases (at 0°C, 1 atm)	Density (g/l)	C_L (km/s)
Acetone	0.79	1.174	Air	1.293	0.331
Benzene	0.87	1.295	Ammonia	0.771	0.415
Carbon tetrachloride	1.595	0.926	Argon	1.913	0.330
Castor oil	0.969	1.477	Carbon monoxide	1.25	0.338
Chloroform	1.49	0.987	Carbon dioxide	1.977	0.259
Ethanol	0.79	1.207	Chlorine	3.214	0.206
Ethyl ether	0.713	0.985	Ethane	1.405	0.313
Ethylene glycol	1.113	1.658	Ethylene	1.260	0.317
Glycerine	1.26	1.904	Helium	0.178	0.965
Kerosene	0.81	1.324	Hydrogen	0.0899	1.284
Mercury	13.5	1.45	Hydrogen sulfide	1.539	0.289
Nitrobenzene	*1.20*	*1.463*	Methane	0.7168	0.430
Nitromethane	*1.14*	*1.33*	Neon	0.900	0.435
TNT (molten)	*1.47*	*2.10*	Nitrogen	1.251	0.334
Turpentine	0.88	1.255	Oxygen	1.429	0.316
Water	0.998	1.497	Sulfur dioxide	2.927	0.213

The rate of attenuation changes with the sound frequency, higher frequencies attenuating faster than lower frequencies.

3.3 Shock Waves

A shock wave is caused by a very high-pressure disturbance moving through a material (refer back to Figure 3.1). In the case of a shock, the disturbance is not smooth or continuous, but is discontinuous (Figure 3.6). The pressures involved are much higher than those of sound waves. The shock pressures can stress the material well beyond its elastic limit, and the material does not return to its

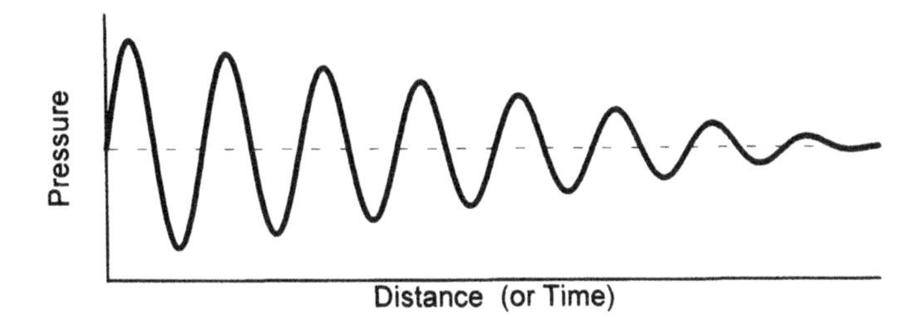

Figure 3.5. Attenuation of a sound wave.

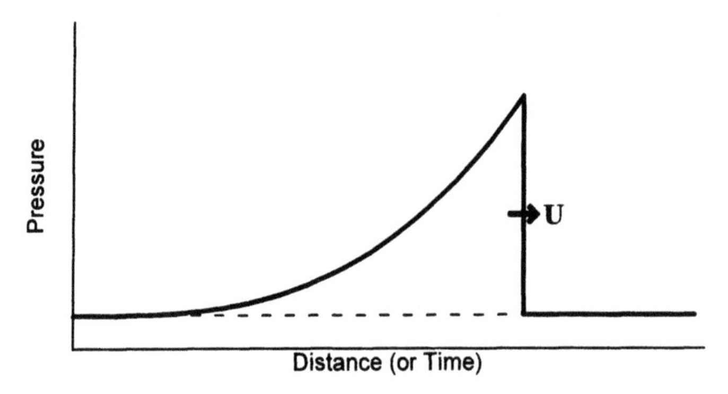

Figure 3.6. Structure of a shock wave.

original state after release of the pressure. Irreversible work has been performed upon it, and this may cause plastic (permanent) deformation of solid materials.

3.3.1 Mass, Momentum, and Energy Balance

As a shock wave moves through a material, the mass, momentum, and energy are conserved across the shock front. This means that the shock process does not create or destroy mass or energy and that the motion that is imparted to particles in front of the shock is caused by the motion of particles bumping into them from just behind the shock. The conservation of these three quantities leads to three equations that describe the shock process. These equations are known as the *Rankine-Hugoniot Jump Equations*.

1. Conservation of mass: $$\frac{\rho_1}{\rho_0} = \frac{U - u_0}{U - u_1}$$

2. Conservation of momentum: $$P_1 - P_0 = \rho_0(u_1 - u_0)(U - u_0)$$

3. Conservation of energy: $$e_1 - e_0 = \frac{P_1 u_1 - P_0 u_0}{\rho_0(U - u_0)} - \frac{1}{2}(u_1^2 - u_0^2)$$

where P is the shock pressure (GPa), U is the shock velocity (km/s), u is the particle velocity (km/s), ρ is the density (g/cm^3), and e is the internal energy.

The subscripts 0 and 1 in the jump equations refer to the state of the unshocked material and the shocked material, respectively. If the unshocked material was at rest and at some low ambient pressure prior to the arrival of the shock, then $u_0 = 0$ and P_0 is so small that it can be considered to be equal 0, and the equations are much simplified.

1. Mass: $$\frac{\rho_1}{\rho_0} = \frac{U}{U - u_1}$$

2. Momentum: $P = \rho_0 u_1 U$

3. Energy: $e_1 - e_0 = \frac{1}{2} P\left(\frac{1}{\rho_0} - \frac{1}{\rho_1}\right)$

3.3.2 The Velocity Hugoniot

The shock velocity is greater than the sound velocity in the unshocked material. It is related to the particle velocity by the *empirical* relationship called the *velocity Hugoniot* equation

$$U = c_0 + su$$

where c_0 is the bulk sound speed, and s the velocity coefficient. The quantities c_0 and s are determined experimentally for each particular material. Table 3.4 gives ρ, c_0, and s for various materials including unreacted explosives.

Example 3.1

In an experiment, we measured the particle velocity to be $u = 1.85$ km/s as a shock wave passed through an aluminum sample. From this measurement alone, along with knowing the unshocked density of the aluminum, we are able to calculate the shock velocity and the shock pressure. What are they?

Solution

a. Known: $u = 1.85$ km/s (measured), the momentum jump equation $P = \rho_0 uU$ (the aluminum was at rest prior to being shocked), the velocity Hugoniot $U = c_0 + su$, and from Table 3.4, for aluminum: $\rho_0 = 2.785$ g/cm^3, $c_0 = 5.328$ km/s, $s = 1.338$.
b. Find the shock velocity first:

$$U = c_0 + su$$

$$U = 5.328 \text{ km/s} + (1.338)(1.85 \text{ km/s}) = 7.803 \text{ km/s}$$

c. Now find the pressure:

$$P = \rho_0\, u\, U$$

$$P = (2.785 \text{ g/cm}^3)(1.85 \text{ km/s})(7.80 \text{ km/s}) = 40.2 \text{ GPa}$$

3.3.3 Attenuation of Shock Waves

Shock waves also attenuate as they travel through a material, although the process of attenuation is somewhat different from that in a sound wave. The shock wave not only loses energy as it travels through a material, but it also loses pressure due to a rarefaction wave that overtakes it. To understand this, consider Figure 3.7, depicting a square-wave pulse. The front of the shock is traveling at a velocity U, which is determined by P and ρ_0. The rarefaction wave in the shock-compressed material is traveling at a velocity, R, determined by the quantities, ρ, P, and u of the material behind and in front of it.

Table 3.4 Shock Velocities for Various Materials

Material	Density (g/cm^3)	Bulk Sound Speed c_0 (km/s)	Velocity Coefficient s
2024 Aluminum	2.785	5.328	1.338
304 Stainless Steel	7.896	4.569	1.490
Ammonium nitrate	*0.86*	*0.84*	*1.42*
Baratol	*2.63*	*2.79*	*1.25*
Beryllium	1.851	7.998	1.124
Brass	8.450	3.726	1.434
Cadmium	8.639	2.434	1.684
Comp B	*1.70*	*2.95*	*1.58*
Copper	8.930	3.930	1.489
Cyclotol (75/25)	*1.729*	*2.02*	*2.36*
Epoxy resin	1.186	2.730	1.493
Gold	19.240	3.056	1.572
Iron	7.850	3.574	1.920
Lead	11.350	2.051	1.460
Lucite	1.181	2.260	1.816
Magnesium	1.740	4.492	1.263
Mercury	13.540	1.490	2.047
Neoprene	1.439	2.785	1.419
Nickel	18.874	4.602	1.437
Nitromethane	*1.13*	*2.00*	*1.38*
Nylon	1.140	2.570	1.849
Octol	*1.80*	*3.01*	*1.72*
Paraffin	0.918	2.908	1.560
Pentolite (50/50)	*1.67*	*2.83*	*1.91*
Polyethylene	0.915	2.901	1.481
Polystyrene	1.044	2.746	1.319
Silver	10.490	3.229	1.595
Tantalum	16.654	3.414	1.201
Teflon	2.153	1.841	1.707
Tin	7.287	2.608	1.486
Titanium	4.528	4.877	1.049
TNT (cast)	*1.63*	*2.57*	*1.88*
Tungsten	19.224	4.029	1.237
Water	0.998	1.647	1.921
Zinc	7.138	3.005	1.581

R, however, is traveling into a material at density ρ (it is already shocked and has been compressed), which is greater than ρ_0 into which U is traveling. As a result the rarefaction wave velocity R is greater than the shock wave velocity U and can be expressed as follows:

$$R = c_0 + 2su$$

As the rarefaction wave progresses into the square region, the shock-wave shape changes from a square wave to a form resembling a sawtooth (Figure 3.8). Then

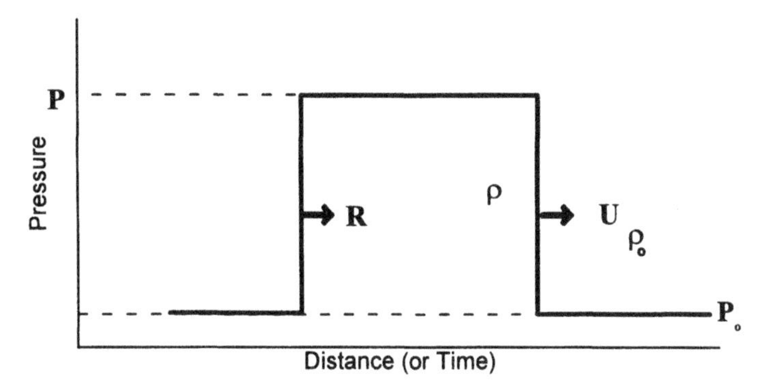

Figure 3.7. Square wave shock pulse.

the rarefaction wave starts to decrease the front pressure. Eventually the peak pressure drops to such a low value that the shock wave becomes a sound wave. As the peak pressure in the front of the shock drops, so does the shock velocity U.

3.3.4 Spallation

If a shock wave in a solid material reaches a free surface of that material, a rarefaction wave starts at the free surface and travels back into the oncoming shock, attenuating it from the front end. This rarefaction is formed because there is no more material ''holding'' against the shock at the free surface.

In other words, the material at the free surface is at a high pressure, and because it is unconfined, it relieves. The free surface moves forward at twice

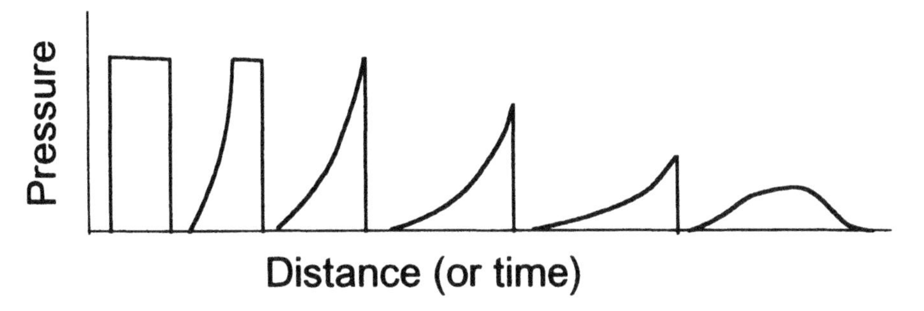

Figure 3.8. Attenuation of a square shock wave.

the particle velocity in the shock. As this reflected rarefaction wave moves farther and farther toward the back of the shocked region, it eventually meets the attenuating rarefaction wave coming up from the back. These two rarefactions have caused the relative particle velocities in the material behind them to be in opposite directions. Therefore, when the rarefactions meet, the material is suddenly pulled into tension at the meeting plane. If the tension is greater than the dynamic tensile strength of the material, then the material will be torn apart at that plane. This process is called spallation.

3.4 Detonation Waves

A detonation wave (Figure 3.9) is a shock wave in a reacting (explosive) material where the chemical reaction is carried out in the shock front. This chemical reaction continuously adds energy to the shock and compensates for energy lost.

When the energy added to the wave front by the chemical reaction is in equilibrium with the energy lost by work and expansion of the rear-flowing product gases, the wave achieves a constant velocity *D*. This is called a "steady-state detonation." The steady-state detonation wave does not attenuate; it retains its structure and detonation state values throughout the detonation process, until all the explosive material is consumed.

The structure of the detonation wave is a bit more complicated than that of a nonreactive shock wave because of the effect of the ongoing exothermic chemical reaction. Referring to Figure 3.9, the front of the wave is traveling at detonation velocity *D*. The spike at the front end, called the Von Neumann spike, initiates the explosive reaction. The reaction zone is where all the chemical reaction takes place and the chemical energy (the heat of detonation) is released.

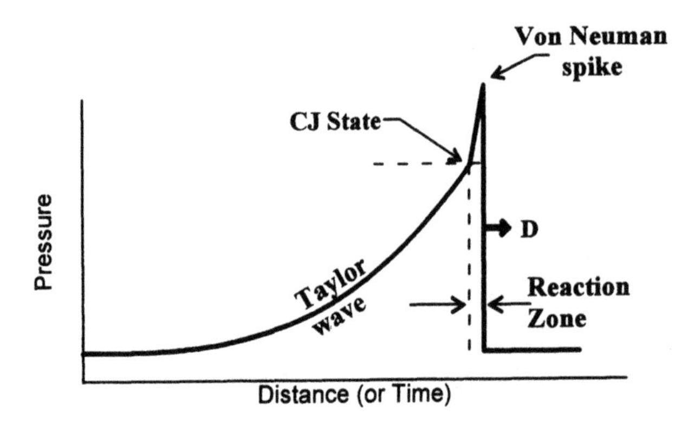

Figure 3.9. Structure of a detonation wave.

The reaction zone in most explosives is approximately 1 mm thick. The end of the reaction zone is called the Chapman-Jouget (C-J) plane, and its state (pressure, density, particle, and shock velocity) is characteristic of the particular explosive material at a given initial unreacted density. Behind the C-J plane, the hot, high-pressure product gases expand in a manner governed by the boundary conditions. The coupling of energy from detonated explosive into another material takes place during this gas expansion. The decrease in pressure due to this expansion is called the Taylor wave.

3.4.1 The Detonation Jump Equations

As in a shock wave, the Rankine-Hugoniot jump equations apply across the detonation front. The shock pressure is replaced by the C-J pressure P_{cj}, the shock particle velocity by u_{cj}, the shock density by ρ_{cj}, and the shock velocity by the detonation velocity D.

Mass balance: $$\frac{\rho_{cj}}{\rho_0} = \frac{D}{D - u_{cj}}, \text{ and}$$

Momentum balance: $$P_{cj} = \rho_0 u_{cj} D,$$

where P_{cj} is the C-J pressure (GPa), ρ_{cj} the density in the C-J state (g/cm^3), ρ_0 the density of the unreacted explosive (g/cm^3), u_{cj} the particle velocity at the C-J state (km/s), and D the detonation velocity (km/s).

The momentum equation can be further simplified by the following approximation, which predicts C-J pressure within an accuracy of 7% for most explosives.

$$P_{cj} = \rho_0 D^2/4$$

Table 3.5 lists the detonation velocity, C-J pressure, and density for some common explosives.

Example 3.2

An explosive has a density 1.70 g/cm^3. We measure its detonation velocity and find it is 7.80 km/s. What is its C-J pressure?

Solution From the approximate momentum equation $P_{cj} = \rho_0 D^2/4$,

$$P_{cj} = (1.70)(7.80^2)/4 = 25.8 \text{ GPa}.$$

3.4.2 Effect of Density

The density affects the detonation velocity. For most explosives, the relationship between detonation velocity D and density of the unreacted explosives ρ_0 is linear. This means it takes the form

$$D = a + b\rho_0$$

Table 3.5 Detonation Properties of Some Common Explosives

Explosive	Density (g/cm^3)	Detonation Velocity (km/s)	C-J Pressure (kbar)	Heat of Detonation (kcal/g)
Baratol	2.550	4.87	140	0.72
Composition B-3	1.715	7.89	287	1.40
Composition C-4	1.590	8.04	257	1.40
HMX	1.890	9.11	390	1.48
HNAB	1.600	7.31	205	1.42
HNS	1.600	6.80	200	1.36
NG	1.590	7.65	253	1.48
PBX-9404	1.840	8.80	375	1.42
PBX-9407	1.600	7.91	287	1.46
Pentolite (50/50)	1.700	7.53	255	1.40
PETN	1.770	8.26	335	1.51
RDX	1.767	8.70	338	1.48
TACOT	1.610	6.53	181	1.35
TATB	1.880	7.76	291	1.08
TETRYL	1.710	7.85	260	1.45
TNT	1.630	6.93	210	1.29
Lead azide	3.800	5.50	299	0.37
Lead syphnate	2.900	5.20	204	0.46

where a and b are empirical constants specific to each particular explosive. Values of a and b for several different explosives are listed in Table 3.6. If there are no data for the velocity-density equation, but D is known at one density, then it can be estimated at another through knowledge of the value of b

$$D_1 - D_2 = b(\rho_1 - \rho_2)$$

When ρ_0 is changed over a small range, it can be assumed that $b = 3[(\text{km/s})/(\text{g/cm}^3)]$. This approximation is normally good to within 10% of the change in D

$$D_1 = D_2 + 3(\rho_1 - \rho_2)$$

Example 3.3

We are going to use the same explosive from Example 3.2 in a system where it will be pressed to a lower density. It will be loaded at 1.2 g/cm^3. What will its detonation velocity be at the lower density? What will its C-J pressure be at the lower density?

Solution Since we have no experimental data to use, we can estimate the changes from the approximation of $b = 3$ in the density-velocity relationship

$$D_1 = D_2 + 3(\rho_1 - \rho_2)$$

$$D_1 = (7.8 \text{ km/s}) + \left(3 \frac{\text{km/s}}{\text{g/cm}^3}\right)(1.2 \text{ g/cm}^3 - 1.7 \text{ g/cm}^3) = 6.3 \text{ km/s}$$

Table 3.6 Density-Velocity Constants for Several Different Explosives

Explosive	a (km/s)	b (km/s)/(g/cm^3)	Density Range (g/cm^3)
Ammonium perchlorate	1.146	2.57	0.55–1.0
BTF	4.265	2.27	—
DATB	2.495	2.83	—
HBX-1	−0.063	4.305	—
LX-04	1.733	3.62	—
Nitroguanidine	1.44	4.015	0.4–1.63
PBX 9010	2.843	3.10	—
PBX 9205	2.41	3.44	—
PBX 9494	2.176	3.60	—
PETN	2.14	2.84	<0.37
PETN	1.821	3.7	0.37–1.65
PETN	2.888	3.05	>1.65
Picric acid	2.21	3.045	—
RDX	2.56	3.47	>1.0
TATB	0.343	3.47	>1.2
TNT	1.67	3.342	—

And now calculating P_{cj} from the detonation momentum balance equation

$$P_{cj} = (1.2\ \text{g/cm}^3)(6.3\ \text{km/s})^2/4 = 11.9\ \text{GPa}$$

3.4.3 Effect of Diameter

As explosives detonate, energy is lost toward the rear where the gases are expanding backward. In the case of steady-state detonation, as mentioned earlier, this is balanced by the addition of chemical energy at the wave front. When the explosive is in the form of a cylinder and it detonates along the axis, the gases also can expand out sideways. The side losses of energy are negligible in large-diameter charges. However, as the diameter is decreased, the side losses become relatively more important, and a drop in the detonation velocity becomes more and more apparent until a diameter is reached where detonation fails to propagate at all. This diameter is called the *critical* or *failure diameter* d_c. The detonation velocity drops linearly with the reciprocal of the diameter and then drops rapidly as the failure diameter is approached. The data for most explosives plotted in this manner do not veer off linearity until the diameter approaches within less than 150 percent of the critical diameter. This is shown in Figure 3.10, where data for Composition B are shown plotted as detonation velocity versus reciprocal diameter.

When the linear portion of the data as represented in this manner is extrapolated to the axis where the reciprocal diameter equals zero (diameter equals

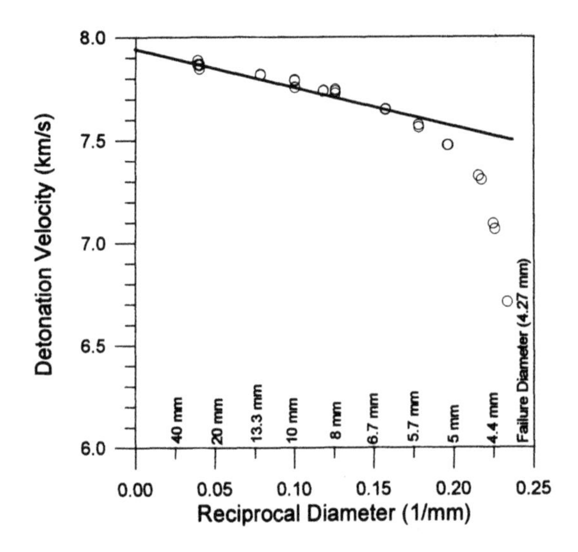

Figure 3.10. Detonation velocity of Comp-B explosive plotted versus the reciprocal of the charge diameter.

infinity), a detonation velocity is obtained that is called the *detonation velocity at infinite diameter* D_∞. If the velocity-diameter data are now replotted as the measured detonation velocity D divided by D_∞ versus the reciprocal diameter, then the obtained fits for all explosives will intercept the velocity axis at $D/D_\infty = 1$. This is shown in Figure 3.11, where data in this form are plotted for four different explosives.

The data from Figure 3.11 fit the simple linear equation

$$\frac{D}{D_\infty} = 1 - A_d \frac{1}{d}$$

where A_d is called the velocity-diameter coefficient and d is the charge diameter. Table 3.7 lists values of A_d, D_∞, and d_c for a number of explosives.

The failure diameter is also dependent on other obvious physical parameters besides the chemistry. Major effects on d_c are caused by confinement, density, particle size, initial temperature, and the presence of additives.

3.5 Explosive Output Tests

The output of explosives can be classified according to C-J pressure, detonation velocity, or heat of detonation. The output is also classified *empirically* by various system tests that measure such nebulous quantities as "brisance," "relative blast," or "relative effectiveness." These tests measure output performance in

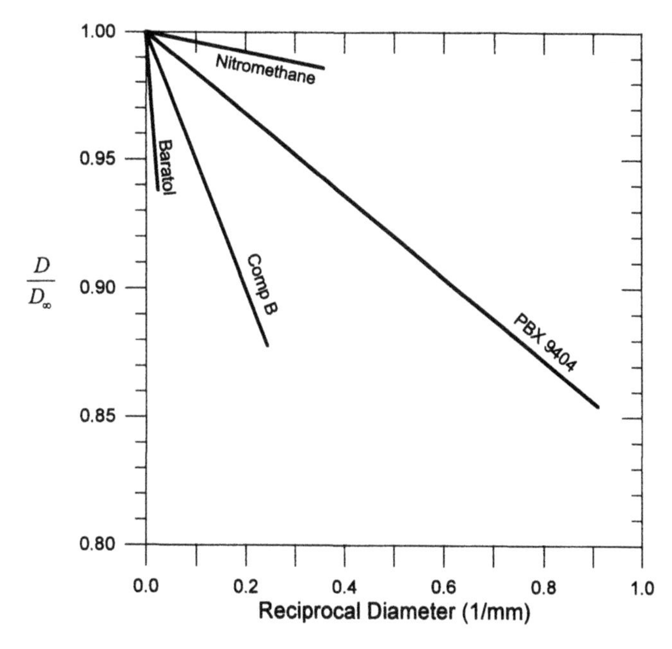

Figure 3.11. Reduced detonation velocity plotted versus the reciprocal of charge diameter.

Table 3.7 Detonation Velocity-Diameter Data for Several Explosives

Explosive	Density (g/cm^3)	D_∞ (km/s)	A_d (mm)	Critical Diameter (mm)
Baratol	2.620	4.96	1.83	43.20
Composition B-3	1.700	7.99	0.189	4.0
Composition C-4	1.530			3.81
Nitromethane	1.128	6.213	0.26	2.84
HMX/WAX (90/10)	1.100			6.50
Lead azide	3.140			0.50
PBX 9501	1.830	8.80	0.19	1.50
PBX 9502	1.890			9.00
PBX 9404	1.846	8.80	0.0548	1.20
Pentolite (50/50)	1.700			6.70
RDX (low density)	0.900			5.20
RDX/WAX (95/5)	1.050			4.50
TACOT	1.450			3.00
TATB	1.876	7.79	0.0431	6.40
TNT (cast, cloudy)	1.61			12.25
TNT (cast, clear)	1.625			3.50
XTX 8003	1.53	7.26	0.00832	0.36
XTX 8004	1.530			1.40

specific applications. "Brisance" is a French term used to describe the shattering ability of an explosive and relates to the C-J pressure. The "relative blast" or "relative effectiveness" relates to the power of an explosive. For example, in a rock quarry blast, the size of the shattered rock relates to the explosive brisance, and the amount of rock moved relates to the explosive power. The following section will describe several standard empirical tests in use today throughout the world. Some of these tests originated as far back as the turn of the century, when they were all that a designer used. The first five tests described are used by the military in categorizing and comparing explosive output.

3.5.1 Sand Test

The sand test measures the explosive output in terms of quantity of sand an explosive will crush inside a heavy, thick-walled confining bomb. The sand test equipment is seen in Figure 3.12. A charge of 0.4 g of explosive is pressed at 3000 psi into a No. 6 blasting cap cup; 0.25 g of tetryl and 0.20 g of lead azide are then pressed at 3000 psi on top of the test explosive as an initiating charge, which is ignited with a 9-in. length of safety fuse. The sample is placed in the test bomb, which contains 200 g of Ottawa sand. The coarseness of the sand will not allow it to pass through a 30-mesh screen. After the test explosive is detonated, the sand is rescreened, and the amount of sand that now passes through the 30-mesh screen is weighed. The results are reported in grams of

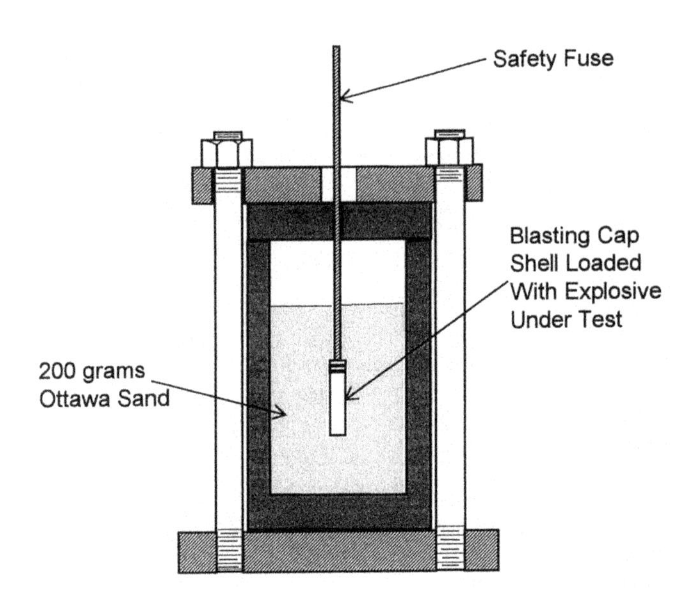

Figure 3.12. Equipment used for a sand test.

sand. The higher the number, the greater the brisance or shattering power of the explosive under test.

The sand test is also used to determine the brisance of liquid explosives. The same procedures are used as in the solid explosive test except for loading the No. 6 cap. The closed end of a No. 6 blasting cap is cut off and loaded with the lead azide and tetryl. Then a 9-in. length of safety fuse is crimped into the cup against the lead azide with the tetryl pressed flush against the end of the cap. The 0.4 g of test explosive is then placed into an aluminum cup with the initiation cap against it, and this assembly is inserted into the test bomb. Sand test values for a number of explosives are listed in Table 3.8.

3.5.2 Ballistic Mortar Test

In the ballistic mortar test, a sample of explosive is loaded into a cylindrical hole in a heavy steel mortar (Figure 3.13). The hole is then plugged with a steel cylinder, the projectile. The mortar is hung from an arm in such a way that it is the ''bob'' on the end of a pendulum. When the test explosive is fired, it propels the projectile out into a pile of sand or dirt and simultaneously kicks the mortar back.

The maximum height (actually, angle) the mortar reaches along the quadrant is recorded. The test is repeated with various weights of text explosive. The quantity of explosive (W, in grams) that causes the mortar to reach the same height that was measured using 10 g of TNT is determined. This result is then reported as ''percent TNT equivalent'' from the following formula:

$$\text{Percent TNT equivalent} = \frac{10 \text{ grams of TNT}}{W} \times 100\%$$

where W is the weight in grams of the test explosive. Values of TNT equivalence as determined by the ballistic mortar test for a number of explosives are listed in Table 3.8.

3.5.3 Trauzl Test

In the Trauzl test, 10 g of the test explosive is loaded into a cavity in a lead cylinder. See Figure 3.14. The cylinder is 200 mm in diameter by 200 mm high. The cavity is centered on the upper face and is 25 mm in diameter by 125 mm deep. After the explosive is fired, the expanded volume of the cavity is measured. This is compared to the cavity volume of that made by a standard 10-g TNT charge.

The results are reported as ''percent TNT equivalent.'' Although the Trauzl test is now seldom used, we describe it because many of the old explosive reference books give Trauzl results. Trauzl test results for a number of explosives are listed in Table 3.8.

Table 3.8 Explosive Output Test Data

Explosive	Sand Crush Sand (g)	Ballistic Mortar (%TNT)	Trauzl (%TNT)	Plate Dent (%TNT)	Fragmentation		Heat of Explosion (cal/g)
					90 mm (No. of Fragments)	3 in. (No. of Fragments)	
Amatol (80/20)	35.5	130	123				490
Amatol (60/40)	41.5	128		583		408	633
Amatol (50/50)	42.5	124		52	630	385	703
Ammonal	47.8	122				550	
AN (neat)	nil	56					346
Baratol (67/33)	26.8			61			740
Baronal	39.8	96					
Black powder	8	50	10				684
Comp a-3	51.5	135		126	1138	710	1580
Comp b	54	133	130	132	998	701	1240
Comp c	46.5	120		112			
Comp c-2	47.5	126		111			
Comp c-3	53.1	126	117	118	944	671	1450
Comp c-4	55.7	130		115			1590
RDX	60.2	150	157	135			1280
Cyclotol (70/30)	56.6	135		136	1165	828	1213
Cyclotol (65/35)	55.4	134			1153	769	1205
Cyclotol (60/40)	54.6	133		132	998	701	1195
DBX	58.5	146		102			1700
DATNB	46.6						
DDNP	47.5	97					820
DEGN	42.2	90	77				841
DNT	19.3	71	64				
LVD dynamite	40.5	92					625
MVD dynamite	52.6	122					935
Ednatol (55/45)	49.4	119	120	112	902	536	1336
Explosive d	39.5	99		91	649	508	800
H-6	49.5	135			714		923
EDNA	52.3	139	122	122		600	1276
HBX-1	48.1	133			910		1840
HBX-3	44.9	111				476	2110
HMX	60.4	150	145				1356
HTA-3	61.3	120					1190
Lead azide	19		39				367
Lead styphnate	24		40				457
Nitromannite	68.5		172				1458
Mercury fulminate	23.4		51				427
Minol-2		143	165	66			1620
NC (12.6%n)	45						855
NC (13.45%n)	49	125					965
NC (14.14%n)	52.3						1058
Nitroglycerine	51.5	140	181				1600
Nitroguanidine	36	104	101	95			721
Octol (70/30)	58.4	115					1074

(Continued)

Table 3.8 *(Continued)*

Explosive	Sand Crush Sand (g)	Ballistic Mortar (%TNT)	Trauzl (%TNT)	Plate Dent (%TNT)	Fragmentation: 90 mm (No. of Fragments)	Fragmentation: 3 in. (No. of Fragments)	Heat of Explosion (cal/g)
Octol (75/25)	62.1	116					1131
Pentolite (50/50)	55.6	126	122		968	650	1220
PETN	62.7	145	173	129			1385
Picramide	48.1	100	107				564
Picratol (52/48)	45	100		100	769	487	1093
Picric acid (presd)	48.5	112	101	107			1000
PIPE	41.6			76	519	428	
Plumbatol	32.4						
PTX-1	54.8	132		127	999	685	1523
PTX-2	56.9	138		141	1128	750	1564
RIPE	40.1	118		85	592	501	
Silver azide	18.9		88				452
Tetracene	28		61				658
Tetryl	54.2	130	125	115	864	605	1105
Tetrytol (75/25)	53.7	122		118	857	591	1099
Tetrytol (70/30)	53.2	120		117	840	585	1098
TNT	48	100	100	100	703	514	1080
Torpex	59.5	138	164	120	891	647	1800
TATB	42.9						1200
Tritonal (80/20) cast		124	125	93	616	485	1770

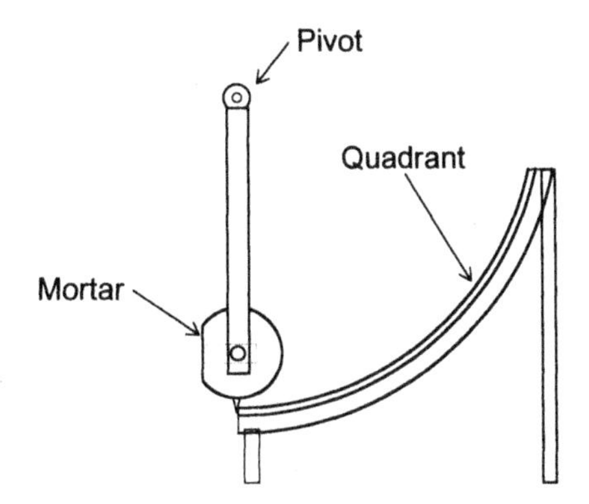

Figure 3.13. Ballistic mortar test.

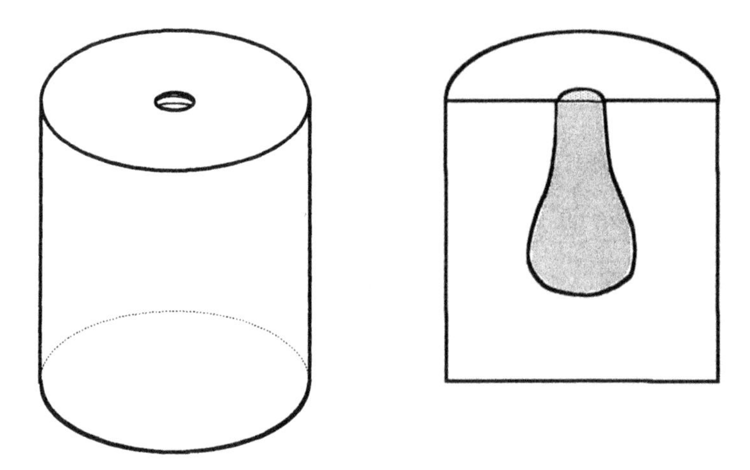

Figure 3.14. Trauzl test.

3.5.4 Plate Dent Test

This is a test in which a charge of a given size is placed on a thick steel plate. After the shot, the depth of dent in the plate is measured and compared to a plate dent caused by the same weight of TNT. Two different dent tests are used that differ in the thickness of plate, size of charge, and manner of supporting the plate. These tests are identified in the reference literature as "plate dent test, method A or B." The results, similar to those of the Trauzl and ballistic mortar tests, are reported as "percent TNT equivalent."

Dent plate tests are used throughout the explosives industry. One such test is used to test production fire sets. A detonator is placed against a steel plate to verify that the fire set had the proper output.

Another popular use of dent tests is for lot testing of explosives. A quantity of explosive from a new batch is fired against a steel plate and its dent compared

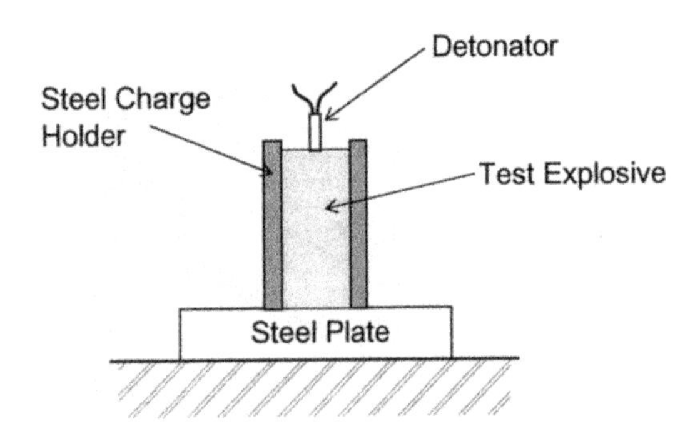

Figure 3.15. Configuration of a plate dent test.

to a standard explosive of the same type. Figure 3.15 shows a typical setup for dent testing. Figure 3.16 is a comparison of dent depths to C-J pressure for various explosives. Table 3.8 lists military dent test (type B) results for a large number of different explosives.

3.5.5 Fragmentation Test

The fragmentation test compares the ability of the test explosive to form fragments from a standard bomb casing. A given quantity of the explosive is loaded into the bomb, and the bomb is buried 4 feet deep in a sand catcher pit. After detonation, the sand in the pit is screened through four-mesh screening. All the fragments retained on the screen are counted. The results are reported as "number of fragments." The object of this test is to determine which explosive will produce the largest number of lethal fragments and not the smallest fragments. There are two standard fragmentation tests that differ only in the explosive load and the bomb used. In one test a 90-mm projectile is used, in the other a 3-in. projectile. Some special tests use other projectile casings such as a 155 mm. Figure 3.17 shows successive screenings of fragments from a test using a 155-mm projectile. Results, reported as number of fragments screened, from the standard 90-mm and 3-in. projectile tests are listed in Table 3.8 for a number of explosives.

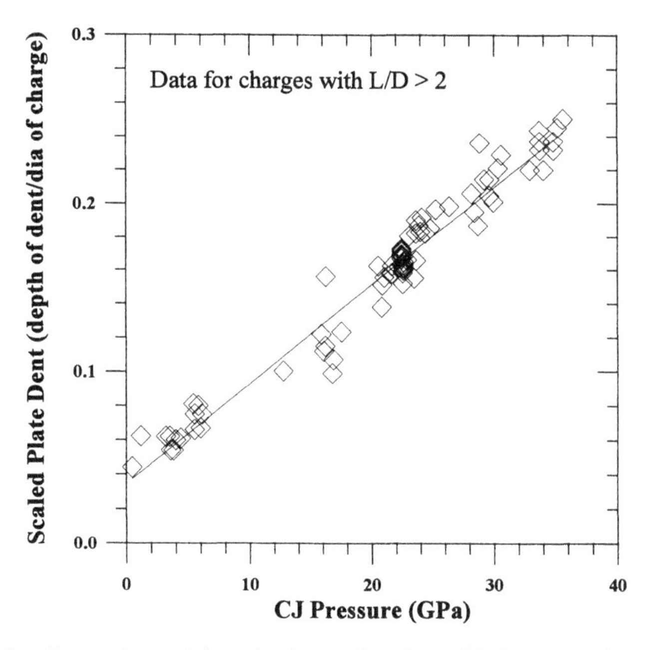

Figure 3.16. Comparison of dent depth as a function of C-J pressure for 34 different explosives.

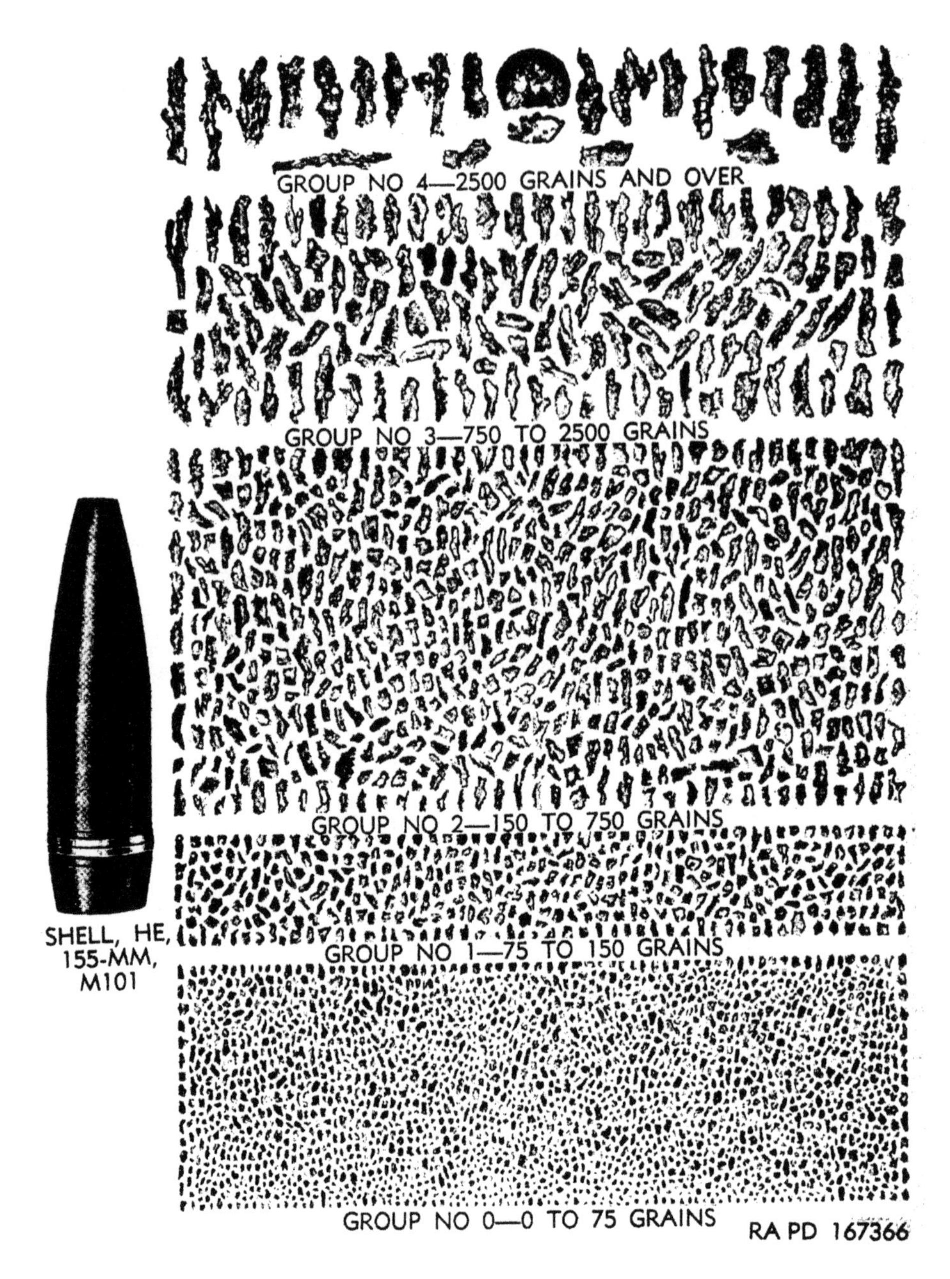

Figure 3.17. Results from a fragment test.

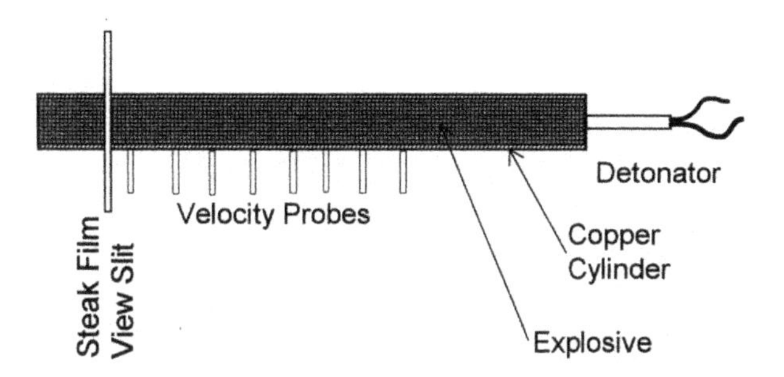

Figure 3.18. Schematic of cylinder test set-up.

3.5.6 Cylinder Test

In addition to the five tests described above, the U.S. Department of Energy uses the cylinder test to measure output. In this test a cylinder of explosive, 1 in. diameter by 12 inches long, is pressed into a copper cylindrical sleeve with a wall thickness of 0.10 in. A high-speed streak camera is set up to record the radial velocity of the copper sleeve as it expands following detonation. (See Figure 3.18.).

The camera is set to measure the expansion at the plane 4 in. from the open end of the charge (8 in. from the detonator). Expansion velocity is reported after the 5 to 6 mm of wall displacement and after 19 mm of displacement. Measurements at these zones allow the data to be reduced into a form yielding output energy ''head-on'' and ''tangential'' to the direction of detonation. The results

Table 3.9 Cylinder Test Data for Several Explosives

Explosive	Density (g/cm^3)	E_{cyl} (kJ/g) Head On	Tangential
Composition B-3	1.728	1.01	1.322
Composition C-4	1.601	0.962	1.258
Cyclotol (77/33)	1.754	1.14	1.445
H-6	1.76	0.769	1.066
HMX	1.894	1.410	1.745
Nitromethane	1.14	0.560	0.745
PBX-9404	1.843	1.295	1.620
Pentolite (50/50)	1.696	0.960	1.260
PETN	1.765	1.255	1.575
RDX	1.80		1.600
TATB	1.854	0.874	1.079
TNT	1.63	0.735	0.975
XTX-8003	1.554	0.710	0.950

are reported as relative energy in these two modes, and are then compared to results from known explosives. Data from cylinder tests of several explosives are listed in Table 3.9.

3.6 Related Reading

1. Courant, R. and Friedrichs, K. O. *Supersonic Flow and Shock Waves*, Interscience Publications J. Wiley & Sons, (1948).
2. Cook, M. A. *The Science of High Explosives*, Reinhold Publications (1958).
3. *Explosive Trains*, AMCP 706-179 (March 1965).
4. Zel'dovich, I. A. and Raizier, I. P. *Physics of Shockwaves and High Temperature Hydrodynamic Phenomena*, Academic Press (1967).
5. Johansson, C. H. and Persson, P. A. *Detonics of High Explosives*, Academic Press (1970).
6. *Military Explosives*, TM 9-1300-214/TO 11A-1-34, Departments of the U.S. Army and the U.S. Air Force (Rev. December 1970).
7. *Principles of Explosive Behavior*, AMCP 706-180 (April 1972).
8. Fickett, W. and Davis, W. C. *Detonation*, University of California Press (1979).
9. Mader, C. L. *Numerical Modeling of Detonations*, University of California Press (1979).
10. Gibbs, T. R. and Popolato, A. *LASL Explosive Property Data*, University of California Press (1980).
11. Marsh, S. P. *LASL Shock Hugoniot Data*, University of California Press (1980).
12. Fickett, W. *Introduction to Detonation Theory*, University of California Press (1985).
13. Blazynski, T. Z. *Materials at High Strain Rates*, Elsevier Applied Science Publications (1987).
14. Cheret, R. *Detonation of Condensed Explosives*, Springer-Verlag Publishers (1993).
15. Graham, R. A. *Solids Under High-Pressure Shock Compression*, Springer-Verlag Publishers (1993).
16. Kuhl, A. L. et al., *Dynamics of Detonations and Explosions: Detonations*, American Institute of Aeronautics and Astronautics, Inc. (1991).

CHAPTER

4

Initiation and Initiators

In this chapter, we shall begin by examining various theories of initiation of both burning and detonation and the energy and power limits found. We will see a number of empirical tests that are used to characterize the input sensitivity of explosives. These tests are useful both for safety considerations and design criteria. Following that, we will examine the various types of initiators and detonators. We will see how they are designed, and what limits their use, as well as safety criteria. The initiator types include nonelectric, hot-wire, exploding bridgewire (EBW), and slappers.

4.1 Initiation Theory and Criteria

When we refer to initiation, we mean a process that starts a reaction. We must further specify, then, whether we are initiating a burning reaction or a detonation. In both cases we initiate a combustion reaction, and this reaction is the ignition stimulus. Let us consider first the initiation of burning, sometimes called ignition.

4.1.1 Hot-Spot Theory

If heat is put into a small local volume of reactive material, the material starts to decompose. The higher the temperature, the faster the decomposition reaction proceeds. As this small volume is raised in temperature, it also transfers heat to the surrounding material. If the reaction in the small heated volume does not produce heat faster than heat is being transferred to the adjacent material, then it cools; the reaction slows down and may eventually stop. If the reaction in the small volume produces heat faster than it can be transferred to the adjacent material, then the little volume heats up or increases in temperature. The increase in temperature increases the reaction rate and heat is produced even faster. Thus an accelerating and self-sustaining reaction has been started. This, in turn, spreads to the surrounding material. This is the hot-spot theroy of initiation.

Hot spots are caused in a variety of ways. The material can be heated with a flame, with hot gases, with hot sparks or particles, with a small hot wire, by friction, or by compression heating. In the latter case, compression causes small local areas such as bubbles or gas pockets to heat faster and to a higher temperature than the material around the bubble. In friction, small local inclusions such as grit, which are harder and have a higher melting point and lower heat capacity than the surrounding material, heat up to higher temperatures. Thus we see that in hot-spot initiation, we rely on heating a very small local area very rapidly to cause a self-sustaining reaction.

4.1.2 Deflagration to Detonation

Once a self-sustaining reaction has begun, it propagates through the adjacent material at a rate determined by porosity, particle size, density, and, most important, by pressure. The reasons for this were developed in Chapter 2.

If the ignited material is a highly confined, detonatable explosive and its diameter is greater than its failure diameter, then as reaction product gases are produced, the pressure and temperature rise rapidly and the reaction rate increases. If the colum of explosive is sufficiently long, then eventually the ever-increasing reaction propagation rate will approach shock velocities, and the reaction undergoes an abrupt transition from deflagration to detonation. The deflagration-to-detonation transition, or DDT, depends upon all the factors that govern burning rate: particle size, porosity, bulk density, gas generation rate, and pressure confinement. Some granular secondary explosives, such as CP and PETN, can be made to undergo DDT in less than one-quarter of an inch under ideal or optimum conditions. On the other hand, under the best conditions, cast TNT requires a DDT distance on the order of 1 foot. Very large masses of even very insensitive explosives such as fertilizer-grade ammonium nitrate can undergo DDT. The huge Texas City explosion was attributed to this when a cargo ship in the harbor caught fire and then exploded, destroying the entire waterfront district. Primary explosives, specifically lead and silver azides, undergo DDT so rapidly and in such a short distance that that quantity has not been measurable. Thus, for these latter two explosives, to ignite them is to detonate them instantly. This behavior of primary explosives distinguishes them from secondary explosives. Although DDT may be consciously used to produce detonation, it does present a threat in hazardous circumstances such as weapon impact, or in fires involving stored explosives.

4.1.3 Impact Initiation

In addition to using DDT, explosives can be made to detonate by initiating them with a shock wave. This is how explosives initiate each other across air gaps and component or train interfaces or in a through-bulkhead initiator. When a shock wave enters an explosive, it compresses the explosive, thereby heating it and starting a chemical reaction. At the same time, it imparts a shock and particle velocity to the reacting material. If the initial shock is of high enough pressure P and long enough duration t, then it will initiate detonation in the explosive. Note that two criteria must be met: high enough pressure and long enough duration. The combination of these two factors, along with the initial density of the explosive and the shock velocity imparted by the input pressure, is called the *critical energy fluence* and is expressed as

$$E_c = (P^2 t)/\rho_o U)$$

If the value of E_c, for the incident shock is greater than the critical value for that explosive, the explosive will detonate. The explosive will not detonate if E_c is below the critical value. The value of threshold or critical E_c is different for each

Table 4.1 Critical Energy Fluence

Explosive	Density (g/cm³)	E_c, Critical Energy Fluence (cal/cm²)	$\frac{(GPa)^2(\mu s)}{(g/cm^3)(km/s)}$ [a]
Composition B	1.73	44	1.84
Composition B-3	1.727	33	1.38
DATB	1.676	39	1.63
HNS-1	1.555	>34	<1.42
Lead azide	4.93	0.03	0.00125
LX-04	1.865	26	1.09
LX-09	1.84	23	0.961
Nitromethane	1.13	405	16.9
PBX-9404	1.84	15	0.627
	1.842	15	0.627
PETN	≈1.0	~2	~0.0836
	1.0	2.7	0.113
	≈1.6	~4	~0.167
RDX	1.55	16	0.669
TATB	1.93	226	9.44
	1.762	72–88	3.01–3.68
Tetryl	1.655	10	0.418
TNT (cast)	1.60	100	4.18
	1.62	32	1.34
TNT (pressed)	1.645	34	1.42

[a] Although these units appear cumbersome, they are much easier to use in shock calculations.

particular explosive. It is strongly dependent upon density. Table 4.1 shows E_c, threshold or critical values for several explosives.

If the shock wave imparted to the explosive is the result of impact from a flyer plate, then both the pressure and time duration of the incident shock can be calculated using the Rankine-Hugoniot equation for conservation of momentum combined with the velocity Hugoniot for both the unreacted explosive and the flyer material, and the flyer thickness and velocity. A shock is generated in both materials when the flyer contacts the explosive surface. A shock runs into the explosive, and a shock of the same pressure runs back into the flyer.

If we assume that the explosive is at rest prior to being impacted by the flyer, then the momentum equation combined with the velocity Hugoniot for the shock entering the explosive is

$$P = \rho_{o,HE} u (c_{o,HE} + s_{HE} u)$$

The combination of these same equations for the shock running back in the opposite direction into the flyer (initial velocity of the flyer = u_o) is

$$P = \rho_{o,flr}(u_o - u)[c_{o,flr} + s_{flr}(u_o - u)]$$

Since the pressure at the interface is the same in both materials, the above two equations can be set equal to each other, yielding a fairly simple quadratic, which can then be solved for both the interface particle velocity and the pressure. These values can then be used in the velocity Hugoniot of the flyer to find the shock velocity in the flyer.

The shock pressure is maintained at the interface for the time it takes the shock in the flyer to reach the back surface of the flyer and the relief wave to return from that surface to the flyer/high-explosive interface. Although the relief wave velocity is higher than the shock velocity, this time is nearly equal to two times the flyer thickness divided by the shock velocity in the flyer. This time is the t in the $E_c = P^2 t/\rho_o U$ expression.

Example 4.1

A copper flyer plate 0.25 in. thick strikes a target explosive at high velocity and develops a shock at the interface. How long is this shocked state maintained? The shock velocity in the copper is 4.16 km/s (or 4.16 mm/μs).

Solution The shock pressure is maintained at a constant value for the time it takes the shock to travel back through the flyer plus the time the reflected relief wave takes to return from the back of the flyer to the interface.

$$t = \frac{2x}{U}$$

where x is the flyer thickness.

$$U = \text{shock velocity} = \frac{2(0.25\ \text{in.})(25.4\ \text{mm/in.})}{41.6\ \text{mm/}\mu\text{s}} = 3.05\ \mu\text{s}$$

Example 4.2

In the above example, the impact pressure at the interface of the copper flyer and the explosive (which is PBX 9404 explosive) is 55 kbar (5.5 GPa). What is the energy fluence developed? How does this compare to the critical energy fluence E_c for PBX 9404? Will the explosive detonate promptly? (Density of PBX 9404 is 1.85 g/cm^3, and the shock velocity in unreacted PBX 9404 at 55 kbar is 4.2 km/s.)

Solution

$$\text{E(energy fluence)} = \frac{P^2t}{\rho_o U} = \frac{(5.5\ \text{GPa})^2(3.05\ \mu\text{s})}{(1.84\ \text{g/cm}^3)(4.2\ \text{km/s})}$$

$$= 11.9\ \frac{(\text{GPa})^2(\mu\text{s})}{(\text{g/cm}^3)(\text{km/s})}$$

From Table 4.1 we find that the critical energy fluence for PBX 9404 in these same units is 0.627. Since the developed energy fluence is greater than the critical, the PBX-9404 will detonate promptly.

Although the term "prompt" was used above in describing detonation caused by an incident shock wave, the reaction does require a finite time and shock run distance to develop into a steady-state detonation. The run distance is dependent upon the density of the explosive and the incidence shock pressure. The higher the pressure, the shorter the run distance. This is shown in Figure 4.1, where the

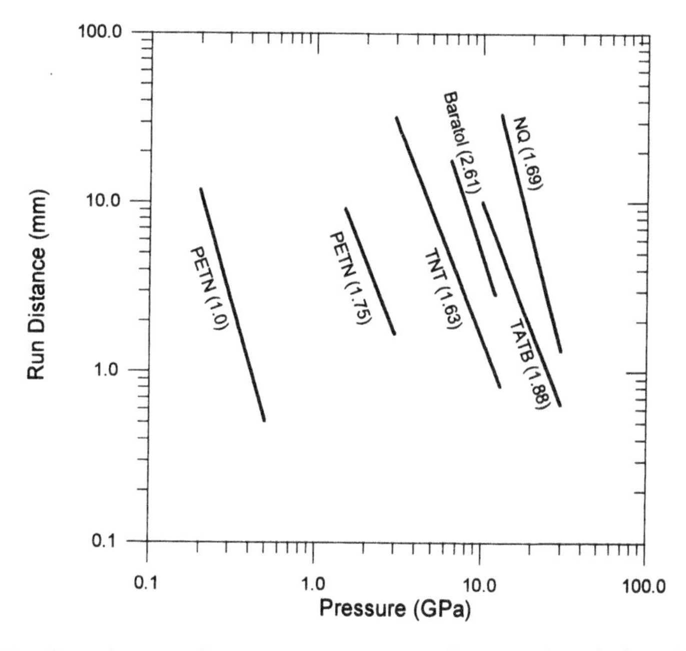

Figure 4.1. Pop plot, run distance versus pressure for several explosives. Density is given in g/cm^3 for each explosive.

log of run distance is plotted versus the log of the incident shock pressure. This particular type of graph is known as a "Pop-plot," named after A. Popolato, formerly of the Los Alamos National Laboratories.

4.2 Initiation Sensitivity Testing

Many empirical tests have been devised to assess the relative safety of handling and manipulating of explosives. These tests attempt to duplicate parts of the complex accident environment to which explosive materials may be exposed. The majority of these tests involve impact in one form or another. There are also tests involving exposure to high-temperature sources and to electrical sparks.

4.2.1 Drop-Weight Tests

These tests consist of dropping a given weight onto a small sample of explosive. The height from which the weight is dropped is varied, and data are recorded as to whether the explosive produced a noticeable reaction or not. In some of the tests, the results are reported as the minimum drop height that produced one sample initiated out of ten samples tested at that height; in others the reported

Table 4.2 Impact Sensitivity for Some Common Explosives

Explosive	Picatinny Machine (in.)	Bureau of Mines Machine (cm)	LANL/LLNL Type-12 (m)	(w/2.5 kg wt) Type-12B (m)
Ammonium nitrate	31	100	1.36	>3.2
Baratol	11	35	0.68–1.4	0.98–1.8
Composition B-3	14	95	0.40–0.80	0.69–1.2
Composition C-4	19	100	0.42	0.36
HMX	9	32	0.32	0.30
HNAB	—	—	0.37	0.32
HNS	—	—	0.54	0.66
Lead azide	3	10	—	—
Lead styphnate	8	17	—	—
Nitroglycerine	7	15	0.20	—
Nitroguanidine	26	47	>3.2	>3.2
PBX-9404	—	—	0.33–0.48	0.35–0.57
PBX-9407	—	—	0.46	0.46
Pentolite (50/50)	12	34	—	—
PETN	6	17	0.13–0.16	0.14–0.20
RDX	8	32	0.28	0.32
TATB	—	—	>3.2	>3.2
Tetryl	8	26	0.37	0.41
TNT	14	95	1.48	1.0

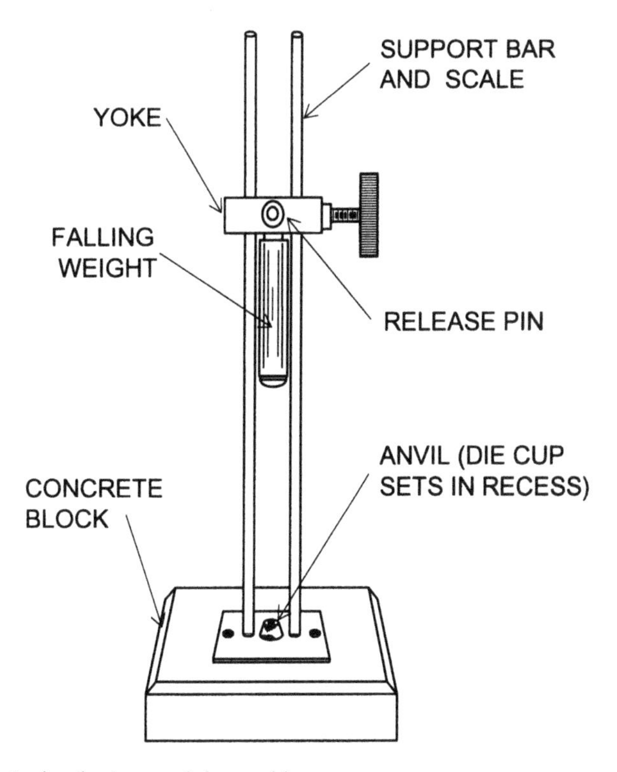

Figure 4.2. A simple drop weight machine

value is the drop height above which 50% of the samples initiate. The various tests differ from each other in the weight and shape of the drop weight, the amount of explosive sample used, and the manner in which the sample is held in the machine (whether it is encased or bare, pressed or loose powder, etc.). The machines that conduct these tests for which most data are available are:

1. Picatinny Arsenal Machine,
2. Bureau of Mines Machine,
3. Los Alamos National Laboratory Machine, and
4. Lawrence Livermore National Laboratory Machine.

The latter two are U.S. Department of Energy tests and are based on the machine design developed in England at the Atomic Weapon Establishment (AWE). They use different anvils, or tools. The type-12 tools have a sandpaper surface, and the type-12B tools have a roughened steel surface. Table 4.2 lists impact sensitivities, as measured by different machines, for several common explosives. Figure 4.2 shows a typical simple drop weight machine.

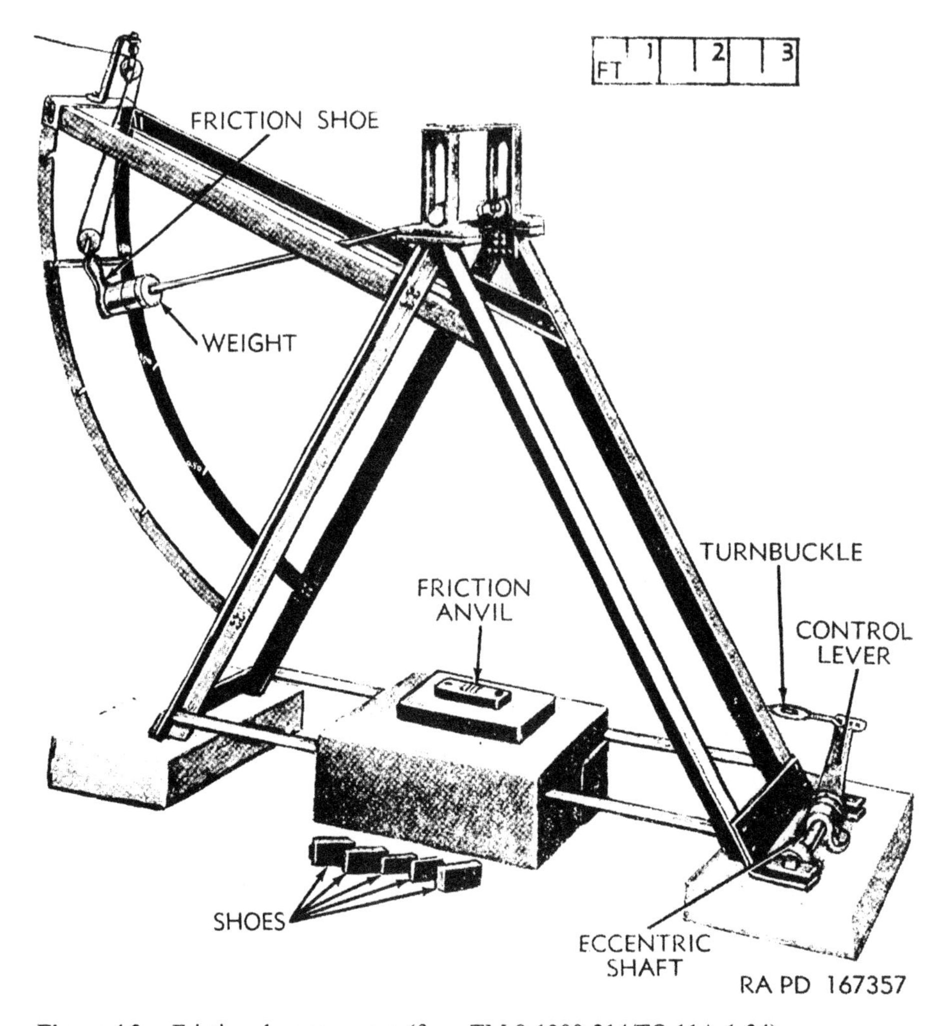

Figure 4.3. Friction shoe apparatus (from TM 9-1300-214/TO 11A-1-34).

4.2.2 Friction Pendulum

In this test, a 20-kilogram shoe is suspended on the end of a pendulum arm. The surface of the shoe is either of steel or fiber. The pendulum is dropped, or swung, from a height of one meter and grazes the surface of a grooved anvil. The test sample of explosive (7 g) is spread evenly over the surface of the grooved anvil. The friction test apparatus is shown in Figure 4.3 The pendulum height is adjusted so that it wipes over the sample 10 times before coming to rest. The behavior of the sample is noted in terms of the severity of its reaction to the friction test. In order of decreasing severity, the terms used to report test

Table 4.3 Friction Pendulum Test Results

Explosive	Steel Shoe	Fiber Shoe
Ammonium nitrate	Unaffected	Unaffected
Composition B-3	Unaffected	Unaffected
Composition C-4	Unaffected	Unaffected
HMX	Explodes	Unaffected
Lead azide	Explodes	Explodes
Lead styphnate	Explodes	Explodes
Nitroglycerine	Explodes	Explodes
Nitroguanidine	Unaffected	Unaffected
Pentolite (50/50)	Unaffected	Unaffected
PETN	Crackles	Unaffected
RDX	Explodes	Unaffected
Tetryl	Crackles	Unaffected
TNT	Unaffected	Unaffected

results are explodes, snaps, crackles, and unaffected. Table 4.3 lists friction pendulum test results for several explosives.

4.2.3 Rifle Bullet Impact

In this test, approximately one-half pound of explosive is loaded into a capped pipe nipple and a bullet is fired through it. The explosive is loaded in the same manner as it would be for service use, that is, cast, pressed, liquid, etc. The inner diameter of the cast iron pipe nipple is 2 in.; the wall thickness is $\frac{1}{16}$ in.; the length is 3 in. Standard cast iron caps are tightly screwed on each end. This arrangement is shown in Figure 4.4.

The pipe bomb is set vertically and a .30 caliber military (ball ammunition) shot is fired perpendicularly into the center of its side from a 30-yard standoff. Five or more tests are conducted, and the results are reported as the percentage that behaved or reacted in each of four categories. The categories, in descending order of violence, are explosion, partial explosion, burned, and unaffected. Table 4.4 lists results of this test for several explosives.

4.2.4 Skid Test

This test was developed at AWE (England) and adapted for use at the Mason and Hanger, Silas Mason, Pantex Plant (Amarillo, Texas) for U.S. Department of Energy explosives, specifically for cast high explosives and PBXs. In this test, 11-in.-diameter hemispheres of the test explosive, weighing 23 pounds, are supported in a cradle at the end of a pendulum device. The pendulum is allowed

Figure 4.4. Pipe nipple and end caps as used in the Rifle Bullet Impact Test (from TM 9-1300-214/TO 11A-1-34).

to swing from preset heights and strikes, at an angle, a sand-coated steel target plate. Two impact angles are employed, defined as the angle between the line of billet travel and the horizontal target surface, 14 and 45 degrees. This pendulum arrangement gives the impact a sliding or skidding component as well as a vertical one and is shown in Figure 4.5.

Table 4.4 Rifle Bullet Test Results

Explosive	Exploded (%)	Partials (%)	Burned (%)	Unaffected (%)
Ammonium nitrate	0	0	0	100
Composition B-3	0	0	0	100
Composition C-4	0	0	20	80
Nitroglycerine	100	0	0	0
Nitroguanidine	0	0	0	100
Pentolite (50/50)	72	20	0	8
PETN	100	0	0	0
RDX	100	0	0	0
Tetryl	13	54	10	23
TNT	40	0	0	60

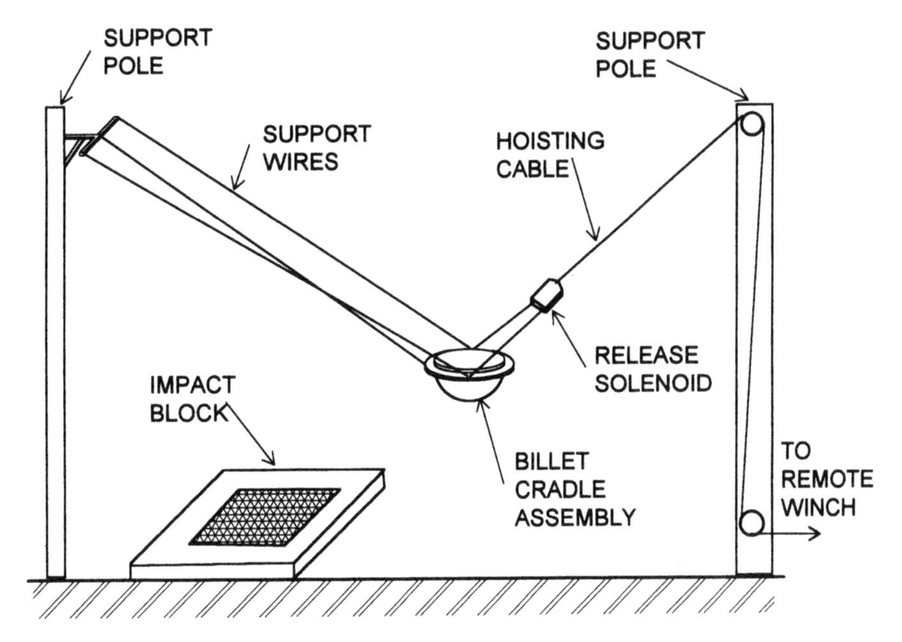

Figure 4.5. Arrangement for the skid test.

Results of the test are expressed in terms of the type of chemical event produced by the impact as a function of impact angle and vertical drop. The chemical events are defined as follows:

0. No reaction; charge retains integrity.
1. Burn or scorch marks on high explosive or target; charge retains integrity.
2. Puff of smoke, but no flame or light visible in high-speed photography. Charge may retain integrity or may be broken into large pieces.
3. Mild low-order reaction with flame or light; charge broken up and scattered.
4. Medium low-order reaction with flame or light; major part of high-explosive consumed.
5. Violent deflagration; virtually all high explosive consumed.
6. Detonation.

A variation of this test, in which all tests are run with an LX-10 explosive billet at a 45-degree impact angle from a 14-foot height, is performed to evaluate various kinds of flooring materials. Table 4.5 shows results of the Standard Skid Test for PBX 9404 explosive on the standard sand-covered steel target.

4.2.5 Susan Test

This is a projectile impact test. The projectile head, or nose, contains approximately 1 pound of explosive, and the target is armor-plate steel. The results are

Table 4.5 Standard LLNL-Pantex Skid Test (Explosive PBX 9404)

Impact Angle (Degrees)	Vertical Drop (Feet)	Chemical Event
88	0.625	0,0,0,0,0,0
14	0.88	0,0,0,0,0,0,0,0,6,0,0,6
14	1.25	0,0,0,6,6,6
14	1.75	6,0,6,0,0,2,0,0,0,6
14	1.90	6
14	2.50	6,0,3,0
14	3.50	6,6
45	1.75	0,0,0,0,0,0,0
45	2.50	0,0,0,0,0,0,0,0,0,0
45	3.50	0,6,0,0,0,0,0,6,0,0
"	"	0,0,0,0,0,0,6
45	5.00	6,6,6,0,6,0,0,6,0,0
45	7.10	6,6
45	10.00	6,6
45	15.00	6

presented as "relative energy release" versus projectile impact velocity. The relative energy release is determined by air shock overpressure measured at a 10-foot distance from the target. On the scale of 0 to 100, complete detonation of a high-energy PBX 9404 would yield a value of 100; complete detonation of cast TNT would yield a value of about 70. Rapid burning reactions, which would consume all the high explosive, would yield a value on this scale of approximately 40. Figure 4.6 shows plots of relative energy release versus projectile impact velocity for a number of shots with 30 different explosives.

The results of Susan tests have been shown to correlate with the response of weapon systems subjected to impact tests. Thus these results are quantitatively useful for more than rank ordering of the impact sensitivities of several main-charge explosives.

4.2.6 Explosion Temperature Test

In this test, a small sample of explosive is pressed into a blasting cap cup made of gilding metal. The cup is then inserted into a molten Wood's Metal bath. The time it takes from insertion into the bath until some noticeable reaction takes place (usually a mild explosion) is then noted. The test is repeated at several different bath temperatures. A smooth curve is drawn through the data points (time to explosion versus bath temperature). The temperature that would cause reaction in 1, 5, and 10 seconds is interpolated from the graph and reported. An additional test is done with this same apparatus, the "0.1 second test." In this

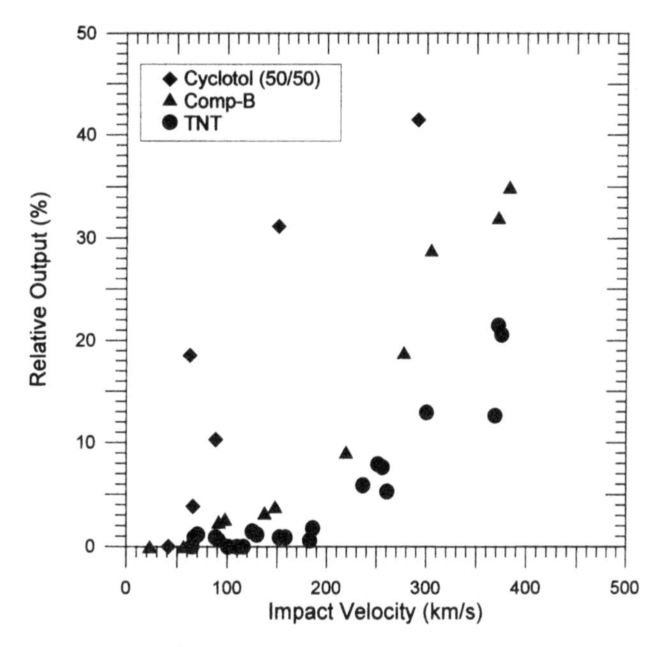

Figure 4.6. Susan test results.

test, a pinch (that is correct, a pinch!) of high explosive is thrown onto the surface of the melted Wood's Metal. If it does not react immediately, then the bath temperature is raised and the test repeated until the temperature is found that causes the thrown sample to "pop" on contact with the bath surface. This is reported as the "explosion temperature at 0.1 second." Table 4.6 gives the results of this test for several explosives.

4.2.7 Los Alamos Skid Test Apparatus

Los Alamos National Laboratory has a combination drop and skid test apparatus for testing the sensitivity of weapon explosives. It consists of a 150-foot tower with two independent hoists that are used to lift the test explosives. The test explosive is dropped on either a 45° incline plane or on a pin depending on which portion of the machine is used. The incline plane has a steel boiler plate 5 inches thick and 4 feet square. A steel plate, covered with 80-grid sandpaper, is placed on the incline for the target, this arrangement is shown in Figure 4.7.

The test explosive is dropped from a predetermined height. If there is no reaction, the drop height is increased until a reaction occurs, then the drop height is lowered until no reaction occurs.

Table 4.6 Temperature (°C) for Explosion Test

Explosive	0.1 s	1 s	5 s	10 s
Ammonium nitrate	—	—	465	—
Baratol	—	—	—	385
Black powder	510	490	427	356
Composition B-3	526	368	278	255
Composition C-4	—	—	290	—
DNT	—	—	310	—
EDNA	265	216	189	178
HMX	380		327	306
Lead azide	396	356	340	335
Lead styphnate	—	—	282	276
Mercury fulminate	263	239	210	199
Nitroglycerine	—	—	222	—
Nitroguanidine	—	—	275	—
Octol (75/25)	—	—	350	—
Pentolite (50/50)	290	266	220	204
PETN	272	244	225	211
RDX	405	316	260	240
Silver azide	310	—	290	—
Tetryl	340	314	257	238
TNT	570	520	475	465

When this "go/no go" height is established, approximately 15 billets are dropped to establish a 50% go/no go. This skid test is called the "down up" method. This apparatus is also used for full-system drop tests and spigot or pin tests. The drops are recorded on high-speed cameras (3000 fps) from two bunkers. Table 4.7 lists a few explosives and their 50% point.

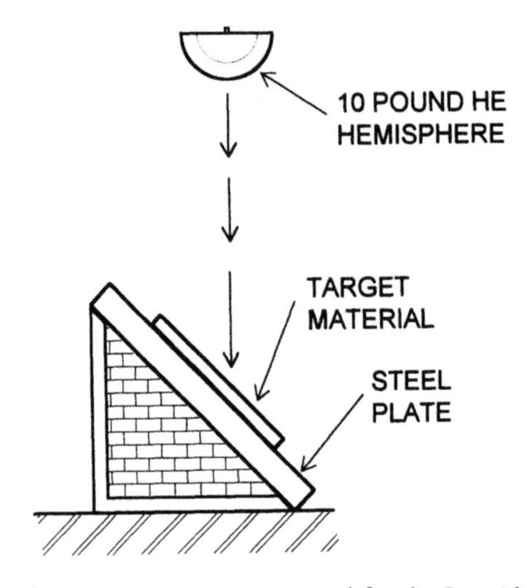

Figure 4.7. Incline-plane target arrangement used for the Los Alamos Skid Test.

Table 4.7 LANL Skid Test Results

Explosive	50% Height (ft) (Go/No Go)
TNT	>150.0
PBX 9010	2.5
PBX 9011	78.0
PBX 9404	4.5
PBX 9501	26.0
PBX 9502	>150.0
LX 09	5.7
LX 10	4.0
Octol	75.0

4.2.8 Gap Tests

There are a number of different "gap tests"; in all of them a fixed charge ("donor") is fired into a thickness of inert material that is in contact, at the output end, with a sample of the test ("acceptor") explosive. In all of the tests the thickness of the gap material is varied such that tests are conducted in which the acceptor does not detonate and does detonate. The "go/no-go" data thus obtained are analyzed to yield a 50% point and statistical standard deviation. The thicker the 50% point gap, the more sensitive the test explosive. In all these tests a dent block is used in contact with the output end of the test sample in order to determine whether or not the sample explosive had detonated. The 50% gap thickness is chosen as the measure of relative shock sensitivity of the test explosive.

a. NSWC small-scale gap test: In this test the donor is a 25.4-mm-diameter × 38.1-mm-long RDX pellet. The gap spacers are 25.4-mm-diameter Lucite™ discs of different thicknesses. The acceptor explosive 25.4 mm-diameter × 38.1 mm long. Results for three typical explosives are: tetryl (1.69 g/cm^3)—7.8 mm, PETN (1.77 g/cm^3)—6.03 mm, Comp B (1.74 g/cm^3)—4.75 mm.
b. LANL small-scale gap test: In this test the donor is a modified SE-1 EBW detonator with an output pellet 7.62 mm diameter × 5.26 mm long of PBX-9407. The gap spacers are brass shims 0.254 mm thick. The acceptor explosive is 12.7 mm diameter × 38.1 mm long. Results for the same three typical explosives are: tetryl (1.69 g/cm^3)—3.84 mm, PETN (1.77 g/cm^3)—5.21 mm, Comp B (1.74 g/cm^3)—0.53 mm.
c. LANL large-scale gap test: In this test the donor is a 41.3-mm-diameter × 101.6-mm-long PBX-9205 pellet. The gap spacers are 41.3-mm-diameter aluminum discs of different thicknesses. The acceptor explosive is 41.3-mm-diameter × 101.6 mm long. Results for two of the above three typical explosives are: tetryl (1.69 g/cm^3)—59.82 mm, and Comp B (1.74 g/cm^3)—44.58 mm.

d. Pantex gap test: In this test the donor is a 25.4-mm-diameter × 38.1-mm-long LX-04 pellet. The gap spacers are 25.4-mm-diameter brass discs in thickness increments of 0.25 mm. The acceptor explosive is 25.4 mm diameter × 25.4 mm long. The resulting 50% gap for Comp B (1.74 g/cm^3) is 23.2 mm.
e. SRI gap test for liquid explosives: In this test the donor is a 41.3-mm-diameter × 25.4-mm-long tetryl pellet. The gap spacers are 41.3-mm-diameter cellulose acetate discs each 0.254 mm thick (referred to as "cards" by the testers). The acceptor explosive is contained in a steel tube (with a wall thickness of 2.54 mm), which has an inside diameter of 12.7 mm, and is 101.6 mm long. Results for three typical explosives are: NG/EGDN (50/50)—45.7 mm, NM—32 mm, NM/TNM (50/50)—40 mm.

4.2.9 Electrostatic Sparks

There are no widespread standard tests yet for testing the electrostatic or spark sensitivity of explosives. A number of tests have been devised, and perhaps a standard will someday be accepted. In general, secondary explosives are not particularly susceptible to this mode of ignition; however, primary explosives and most pyrotechnics are susceptible. Initiation of an explosive material by an electrostatic pulse is achieved if sufficient energy is delivered to the explosive material. Since most explosive materials are not good electrical conductors, the

Table 4.8 Electrostatic Sparks Test Data

	Energy (Joules)		Type of Ignition	
Explosive	Unconfined	Confined	Unconfined	Confined
Black powder	12.5	0.8	none	deflagration
Explosive D[a]	0.025	6.0	deflagration	detonation
Explosive D[b]	12.5	6.0	none	detonation
Lead azide	0.007	0.007	detonation	detonation
Lead styphnate	0.0009	0.0009	detonation	detonation
NC (13.4% N)	0.016	3.1	deflagration	deflagration
Nitroglycerine	12.5	0.9	none	detonation
PETN[a]	0.062	0.21	deflagration	detonation
PETN[b]	11.0	0.21	none	detonation
Tetryl[a]	0.007	4.38	deflagration	detonation
Tetryl[b]	11.0	4.68	none	detonation
TNT[a]	0.062	4.38	deflagration	detonation
TNT[b]	11.0	4.68	none	detonation

[a] Filtered through 100 mesh.
[b] As received.

electrostatic pulse must have sufficient potential to overcome the explosive material's low electrical conductivity.

Federoff and Sheffield, in the *Encyclopedia of Explosives and Related Items*, Volume 5, report the data shown in Table 4.8, but do not describe the apparatus or technique. The data are for the ''highest electrostatic discharge energy at 5000 volts for zero ignition probability of explosive.''

Sandia National Laboratories has developed a uniform test specification for measuring the spark sensitivity of explosive-containing components. A 600-pF capacitor charged to 20 kV is discharged through the component with 500 ohms in series with the discharge circuit. This test is considered to be equivalent (or in excess) of the discharge that might be obtained from a static discharge from the human body. For each component, different test configurations are given, and recently, it has been the policy to state a reliability figure against spark ignition. Spark initiation tests are performed on explosives in the following physical configurations:

4.2.9.1 Loose Powder

The test fixture used for testing bulk powder is shown in Figure 4.8. The spark ignition threshold of the bulk powder is measured by discharging a 600-pF capacitor from a pointed electrode through a sample of loose powder (approximately 0.2 g) to a ground plane.

The voltage is varied until one ignition in ten trials is achieved or until the voltage limit of the test equipment is reached. The 500-ohm series resistor is not used if the powder is resistant to initiation. It is desirable to obtain some ignitions data during the test to allow a comparison of powders. Some results of this test are given in Table 4.9.

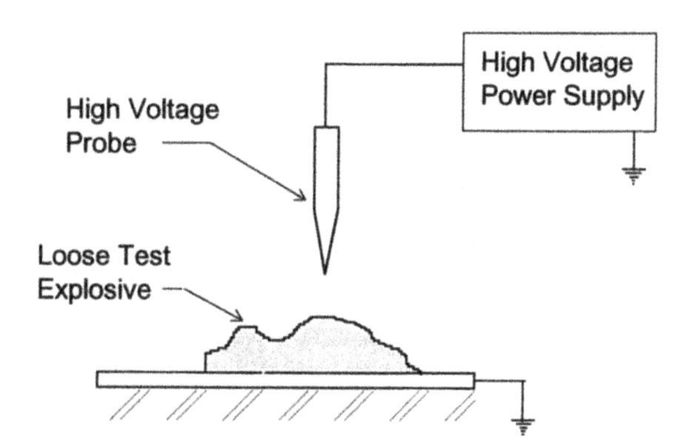

Figure 4.8. Test fixture for spark ignition of loose powder.

Table 4.9 Spark Sensitivity of Loose Powders

Explosive	Capacitance (pF)	Voltage (volts)	Series Res. (ohms)	Ignition Trails
CP	600	35,000	0	0/50
Lead azide	600	5,000	500	5/5
Silver acetylide	600	5,000	500	5/5

4.2.9.2 Compressed Powder

Several different methods were tried to determine similar thresholds for pressed powders that would correspond to the inherent sensitivity of the material as present in the component. The standard component electrostatic sensitivity test (typically a go/no-go test at a preset energy level) does not determine the sensitivity of the powder. It determines merely the efficiency of spark gaps built into the component hardware, either intentionally or inadvertently.

The test fixture that was found to give the most meaningful data for conductive powders is shown in Figure 4.9. The high voltage is applied to the bridge wire electrodes, and the steel ring is grounded. The spark then jumps between the end of the pressed powder column and the ring. Results of this test for PETN and CP at various particle sizes and densities are presented in Table 4.10.

4.3 Nonelectric Initiators

The simplest initiators, simple in mechanical construction but not necessarily in terms of their mechanisms of initiation or of their chemical design or perfor-

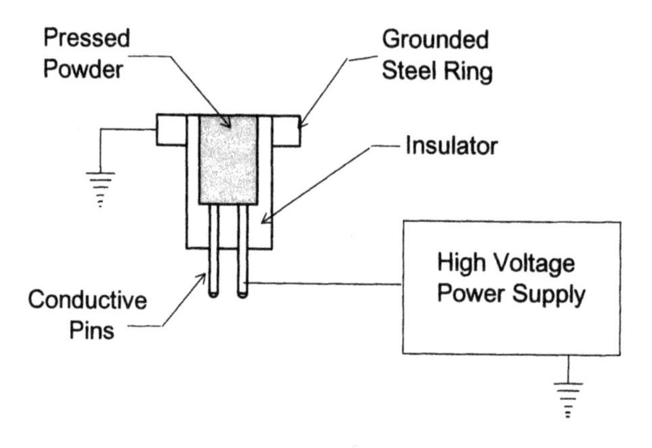

FIgure 4.9. Test fixture for spark ignition of pressed powder.

Table 4.10 Spark Sensitivity of Compressed Powders (All Tests of 600 pF)

Explosive	Density (g/cm^3)	Surface Area (cm^2/g)	Series Resistance (Ohms)	Voltage (Volts)	Ignitions/Trials
PETN	0.88	2,800	0	37,500	1/2
PETN	0.88	4,280	0	20,000	1/10
PETN	0.88	9,044	0	12,500	1/10
PETN	0.88	17,300	0	15,000	1/10
CP	1.20	—	500	35,000	1/10
CP	1.30	—	500	18,800	1/10
CP	1.40	—	500	10,000	1/10
CP	1.51	—	500	10,000	5/10

mance analysis, are the nonelectric initiators. These can be divided into four major categories according to the mechanism of initiation:

1. Flame or spark initiated,
2. Friction initiated,
3. Stab initiated, and
4. Percussion initiated.

4.3.1 Flame or Spark Initiators

These devices are usually detonators. The input end of the detonator has a charge of lead azide that detonates instantly upon exposure to sparks, hot particles, or flame, and a secondary explosive as an output charge. The common nonelectric blasting cap is probably the detonator produced in highest volume. More than one billion (10^9) are manufactured annually in the United States alone. Figure 4.10 shows a typical commercial nonelectric blasting cap. The source of flame and sparks used to initiate this detonator is a safety fuse, or shock tube, which is crimped into the open end.

The lead azide charge has a lacquer seal over it to protect it from moisture. The ''spit'' of the initiator must be strong enough to break or burn through the lacquer seal to initiate the lead azide. The cap housing, or cup, is usually made of copper, gilding metal, or aluminum, but some are made of extrudable steel alloys. The output charge is usually either PETN or RDX. The No. 8 blasting cap is by far the most common. It is loaded with 0.3 g lead azide and 1.2 g output high explosive. The dimensions are 7 mm OD and 40–45 mm long. A No. 6 nonelectrical blasting cap is also manufactured in large quantities, and its total loading is approximately 0.7 g of explosives.

Military fuse trains sometimes use flame-initiated detonators. In the military system they are called ''flash'' detonators. They are usually very small. Figure 4.11 shows two common military flash detonators.

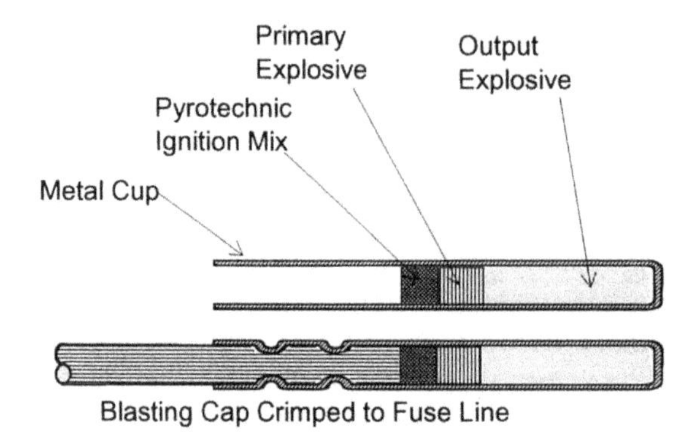

Figure 4.10. Commercial nonelectric blasting cap.

The flash detonators are used in trains preceded by some nondetonating burning element, such as a percussion cap or pyrotechnic delay element. They are used mainly in mechanical out-of-line, safe-and-arm fuse trains and serve the purpose of both explosive relay and detonator.

4.3.2 Friction-Initiated Devices

By far the most common ignitor in the world is the friction ignitor. United States production of one type of inexpensive friction ignitor exceeds 500 billion (5×10^{11}) units annually. This ignitor is called the safety match. The common safety, or book, match operates by bringing into intimate contact two chemicals that then instantly react with each other. Potassium chlorate is mixed with a number of other ingredients, including glue, in the match head. On the striker, the other reactant, red phosphorus, is also mixed with a glue. When the match is rubbed on the striker, the glue coatings on the two reactants are broken and the reactants

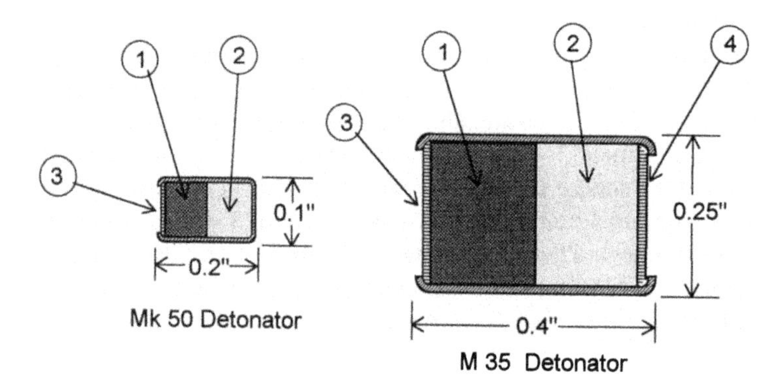

Figure 4.11. Typical military flash detonators. 1. Lead azide, 2. tetryl, 3. input seal, and 4. output seal.

come into intimate contact and immediately react. The friction required to break the coatings is far less than that which can produce high temperature, even with a grit. In the case of the safety match, the mechanism of initiation is really that of hypergols, mixing two chemicals that instantly react with each other. The "strike anywhere" (SAW) match, on the other hand, operates by thermal ignition. Friction raises the temperature of the fine grits that are included. These grits, as hot spots, cause the initiation of burning of the reactants. The major reactants on the tip of the SAW or wooden kitchen match are potassium chlorate and phosphorus sesquisulfide. The grit is powdered glass, carbide, or aluminum oxide. These same materials, as in the SAW match, are also used in the string and tab pull friction ignitors used in some military and commercial systems, as well as in the striker assemblies on fuse flares.

4.3.3 Stab Initiators

This type of initiator is among the most mechanically sensitive of all of the nonelectric types. A typical stab detonator is shown in Figure 4.12. Most stab initiators are fairly similar in size and sensitivity; they use the same standard firing pin (with a few rare exceptions) shown in Figure 4.13.

The firing pin pierces the closure disc and penetrates the priming mix during activation. This causes heating of the mix by both compression of the mix in front of the pin and friction on the mix by the conical sides of the pin. For the same primer mix, at the same load density, the minimum firing energy increases approximately linearly with the thickness of the input closure disc. For the same closure disc thickness, the minimum input firing energy increases with decreasing loading density of the primer mix (Table 4.11).

The firing energy for stab initiators is determined by a drop weight test. A given weight is dropped from various heights onto a centered firing pin, which then pierces the initiator. Numerous tests are conducted at various heights, and the data are reduced to a chart of firing energy (height times weight) versus

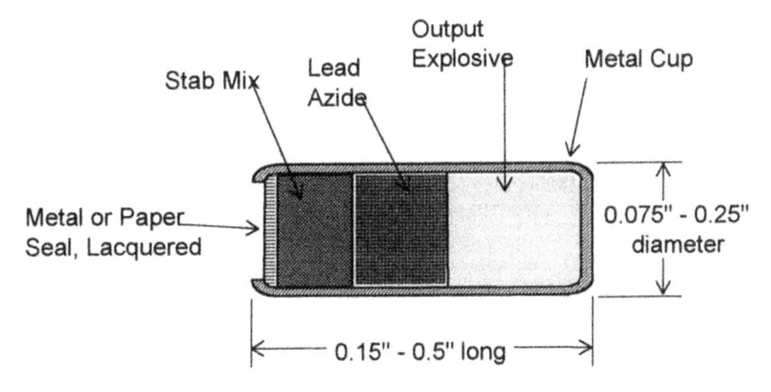

Figure 4.12. Stab detonator.

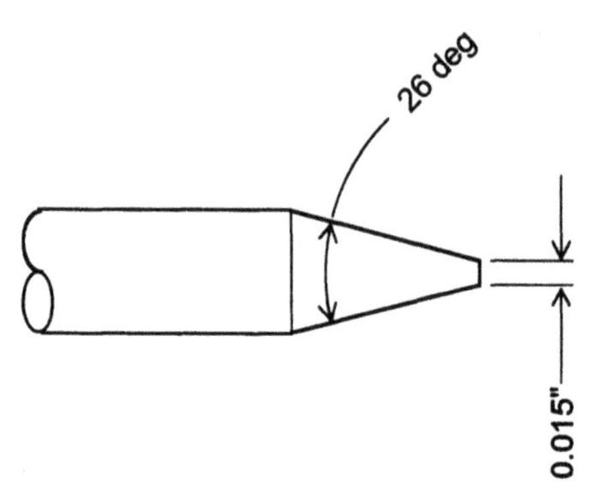

Figure 4.13. Stab initiator firing pin.

probability of function. Most stab initiators function at high reliability at input energy levels between 0.5 and 5 in. ounces (0.0035 to 0.035 joules).

Stab initiators are used in military systems such as small mechanical fuses, where very little mechanical energy is available because of weight limitations and the small dimensions of the springs.

4.3.4 Percussion Initiators

The mechanism for percussion initiation involves only compression of the ignition mix. With these devices the firing pin does not penetrate the case of the primer. Two typical types of percussion primers are shown in Figure 4.14. The firing pins vary with the size of the primer. The striking face of the pin is generally spherical to the full diameter of the firing pin. Small primers, such as pistol-cartridge types, generally require a firing pin with about a 0.05-in. radius.

Percussion primer mixes usually contain a lead styphnate, tetracene, or an antimony sulfide-type mixture (which can act as grit), and a grit to sensitize the mix to crushing and compression. The primer functions when the firing pin

Table 4.11 Effect of Loading Pressure on Stab Initiator Sensitivity (NOL Primer Mix in MK102 Cups, 2 oz Ball Drop Weight)

Loading Pressure (psi)	99% Function Probability Drop Test Height (in.)
93.	1.31
25,000	0.91
40,000	0.77
60,000	0.68
80,000	0.57

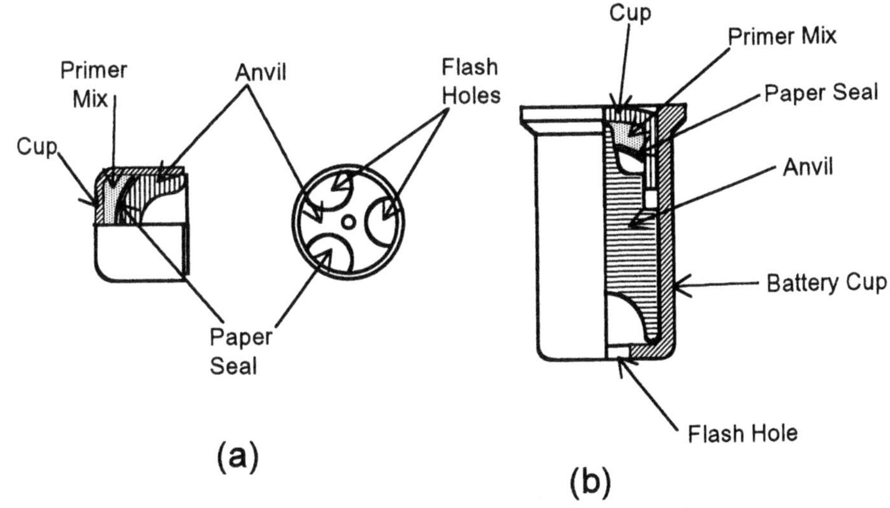

Figure 4.14. (a) Small percussion primer. (b) Large percussion primer.

strikes the cup face and dents it. This crushes or compresses the mix between the inner face of the cup and the anvil. Hot-spot initiation occurs, and flame and sparks are thrown through the ports. As with the stab primers, percussion primers are test initiated with a drop weight apparatus. Small percussion primers have minimum reliable input firing energies in the neighborhood of 20 to 25 in. ounces (0.14 to 0.18 joules). Energy requirements of larger primers may go as high as 150 in. ounces (or a bit over a joule).

Example 4.4

a. A given percussion primer has a listed all-fire energy of 20 in-oz. From what height would a half-ounce firing pin have to be dropped to deliver this energy?

Solution

$$\text{Potential energy} = \text{height} \times \text{weight}$$

$$E = hw$$

$$h = E/w = 20 \text{ in.-oz}/0.5 \text{ oz} = 40 \text{ in.}$$

Example 4.4

b. At what velocity would a 1-oz firing pin have to strike this primer to deliver the required energy?

Solution

$$E = \frac{1}{2}\frac{w}{g}v^2,$$

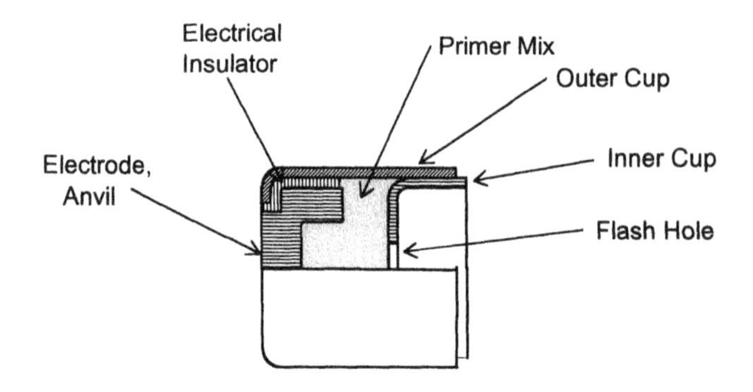

Figure 4.15. Percussion electrical primer.

where E is the kinetic energy, w the weight, g the gravitational constant, and v the pin velocity.

$$v = (2Eg/w)^{1/2} = [2(20 \text{ in.-oz})(32.2 \text{ ft/s})(1 \text{ ft}/12 \text{ in.})/(1 \text{ oz})]^{1/2} = 10.36 \text{ ft/s}$$

Percussion primers are used not only in artillery and small arms cartridges, but also in a wide array of cartridge-actuated devices. A large percentage of the large-bore artillery cartridges use a primer that can be fired either mechanically or electrically. Figure 4.15 shows this primer.

4.4 Hot-Wire Initiators

The vast majority of electric initiators are the hot-wire type. In this type of initiator a small wire is imbedded in, or in contact with, an ignition material, either a primary explosive or a pyrotechnic. Electrical current heats the wire, which in turn heats the ignition material to its ignition temperature and thus starts a burning reaction.

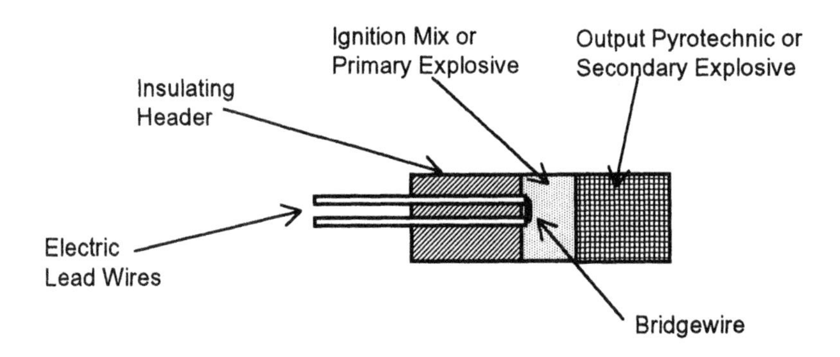

Figure 4.16. Hot-wire initiator assembly.

This reaction is then propagated to the next element in the device, either another pyrotechnic (in the case of an ignitor) or an explosive (in the case of a detonator). The general arrangement is shown in Figure 4.16.

4.4.1 Construction

Although all hot-wire initiators are similar in that they have a header, a bridge, and some kind of explosive material, they differ considerably in size, shape, and construction details. The simplest are the dipped electric matches. This is shown in Figure 4.17.

The electric match has a soldered, or more often, a welded bridgewire. The bridge is "raised"; that is, it is not in contact with the header, which in this case is often a small rectangle of glued cardboard. The ignition charge is buttered or dipped onto the bridgewire; then the solvent in it is evaporated off.

Next, the match is dipped into a slurry of output charge material and dried. Some matches are then further "glazed" by a second dip into a nitrocellulose lacquer to help seal the match from moisture and give the bulb higher mechanical strength.

The next simplest device is probably the electric blasting cap. These caps can be divided, structurally, into two categories: solid-pack blasting caps and match-gap blasting caps. Figure 4.18 shows both of these.

Older electric blasting caps and military electric blasting caps are the solid pack type. More recently produced (early 1970s) commercial electric blasting caps are the match-gap type. The reason is strictly economic. The latter uses a standard nonelectric blasting cap with a rubber header and an electric match crimped into it in lieu of the safety fuse.

The above electric initiators as well as older military detonators, such as the M36A1 (Figure 4.19), all use raised bridges.

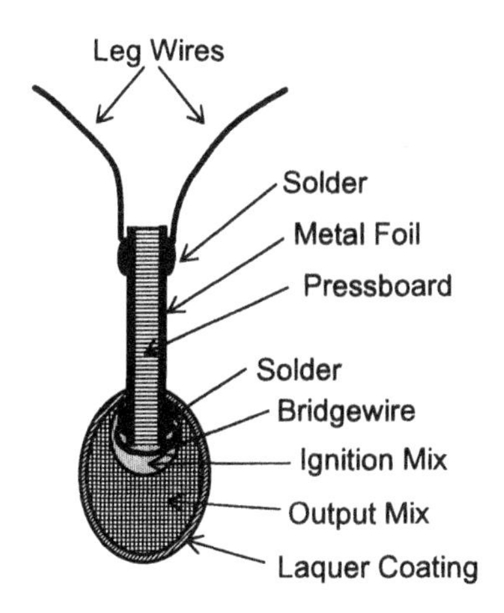

Figure 4.17. Dipped electric matches.

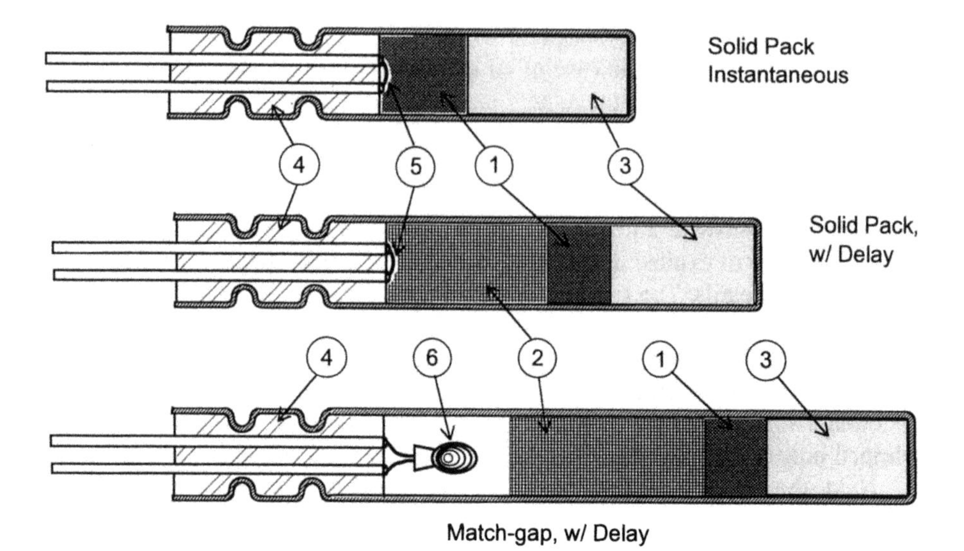

Figure 4.18. Electric blasting caps. 1. Lead azide; 2. pyrotechnics; 3. output secondary explosive; 4. insulating header plug; 5. bridgewire; and 6. electric match.

Other detonators have "flush" bridges, where the bridge is in contact with the header and the ignition material is pressed over it. Flush bridges may be of forms other than metal wires. Older military detonators such as the M48, M51, and T21 used carbon bridges. In some detonators there is no bridge; carbon or graphite or some other electrically conductive material is added to and intimately mixed with the initiating explosive, and the device ignites by heating of the

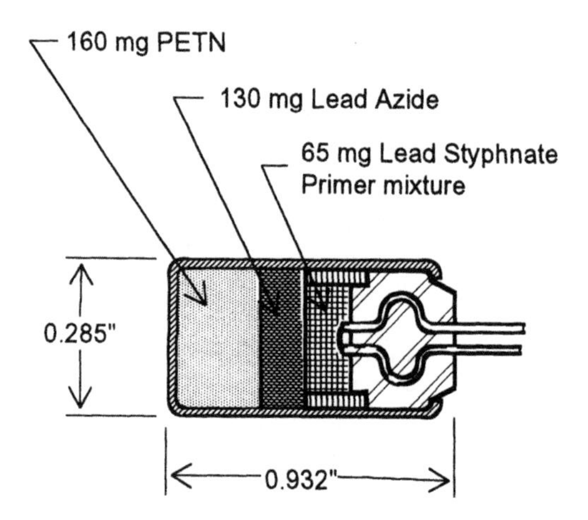

Figure 4.19. M36A1 military detonator.

conductive path through the explosive between two electrodes. Bridges of metal ribbons and vapor-deposited metallic films are used in many newer initiators. The latter give the designer more latitude in controlling heat transfer from the bridge to the header and to the pressed ignition material. Controlling the heat transfer is very important and will be discussed later. The other mechanical feature of electric initiators that should be noted is that they are often incorporated in an electrical connector configuration instead of having lead wires (or pigtails).

4.4.2 Electrical Initiation Criteria

The small metal bridge in a hot-wire initiator acts as the hot spot that was described earlier. When current is applied to the bridgewire, the wire heats up and transfers this energy to the explosive, causing it to react when it reaches its critical temperature.

It is important to differentiate between energy and power. Power is the rate at which energy is being delivered. The total energy delivered to a device is equal to the power (or rate of delivery) times the time during which it was delivered: energy = power × time. In terms of heat, the units of energy are calories or BTUs; in electrical terms, the units of energy are joules or ergs or watt-seconds. The units of power as heat are calories per second or BTUs per second; in electrical terms the units are joules per second or watts.

$$E = pt$$

As energy is applied to the bridge, it is divided between that required to heat up the bridge and that which is lost or transferred to the surrounding materials. The total energy delivered up to the time of ignition when the ignition explosive reaches its critical temperature is given as

$$E = mC_p(T_i - T_o) + \lambda t \tag{4.1}$$

where E is the total energy delivered, m the mass of bridge, C_p the heat capacity of bridge, T_i the ignition temperature, T_o the ambient temperature, λ the average rate of heat transfer over time t to the header and other materials, and t the time.

If the power is very high (that is, energy is delivered very fast), then the bridge heats very quickly to T_i and there is very little time for heat to be dispersed to the surrounding material (t is very small; therefore, λt is very small). Picture the case where power is approaching infinity. Then λt approaches zero, and essentially all the energy goes into just heating the bridge. This represents a minimum energy condition to fire the initiator. The horizontal dashed line in Figure 4.20 represents this condition, where Eq. (4.1) is reduced to $E_m = mC_p(T_i - T_o)$, since λt is zero for this case. No matter how high the power, if less energy is delivered than this amount (t is too short), then the bridge will never reach the initiation temperature T_i, and the device will not fire.

On the other hand, if the power is very low, there is lots of time for heat to

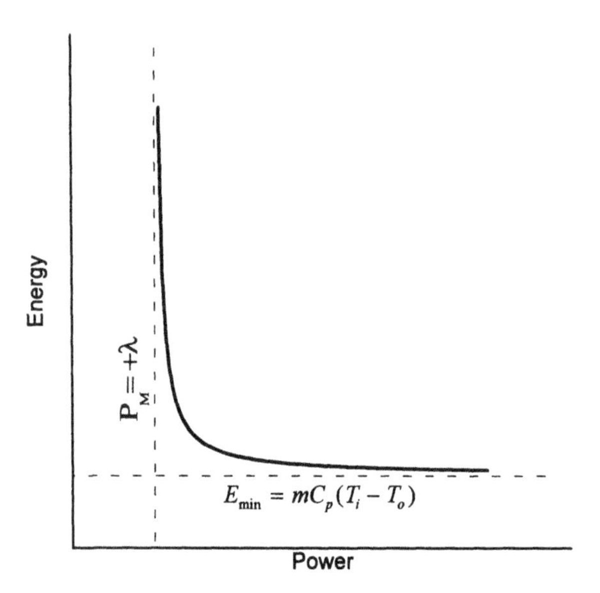

Figure 4.20. Energy-power relationship for ignition and failure.

be lost to the surroundings. If the power is so low that the heat transfer losses dominate, then no matter how long we apply power or how much energy we ultimately deliver, the bridge will not reach T_i, and the device will not fire. This condition is represented by the vertical dashed line in Figure 4.20, where $P_m = +\lambda$, since $mC_p(T_i - T_o)/t$ is zero in this case.

Example 4.5

Suppose we have a hot wire detonator and we know that at room temperature 25°C (298 K) the minimum firing energy is 5.0 millijoules. We also know that the critical ignition temperature of the pyrotechnic mix on the bridgewire is 400°C (673 K).

In order to design a firing system for this detonator for a broad range of environments, we want to know how the minimum firing energy will be affected by temperature extremes. What is the minimum firing energy at −60°F (222 K) and at +160°F (344 K)?

Solution The minimum firing energy was shown to be $E_m = mC_p(T_i - T_o)$. We can assume that mC_p is constant over the range of temperature; so taking the ratio of the energy at one of the extremes, we know that:

$$(E_{m2}/E_{m1}) = [mC_p(T_i - T_2/mC_p(T_i - T_1)],$$
$$E_{m2} = E_{m1}(T_i - T_2)/(T_i - T_1),$$
at $T_2 = -60°F$ (222 K): $E_m = 5(673 - 222)/(673 - 298) = 6.0$ millijoules,
and at $T_2 = 160°F$ (344 K): $E_m = 5\ (673 - 344)/(673 - 298)$
$= 4.4$ millijoules.

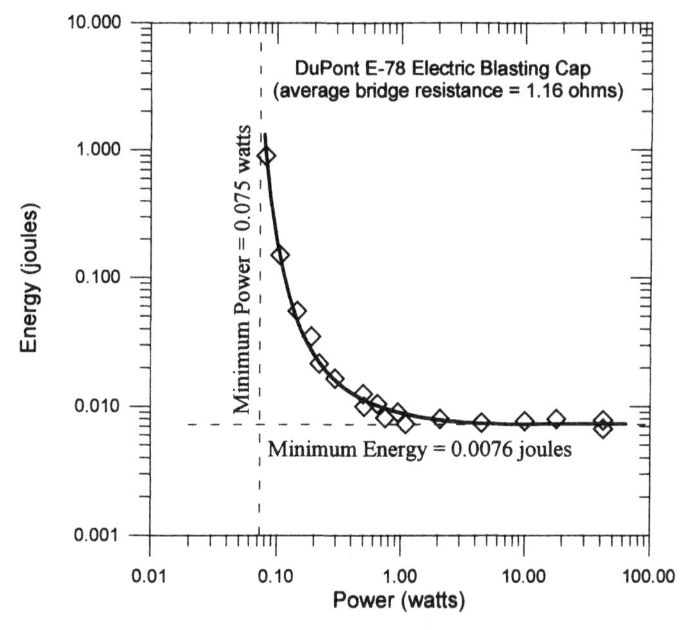

Figure 4.21. Idealized energy-power firing curve for Dupont E-78 blasting cap.

These two limits, minimum power and minimum energy, are real design limits. All hot-wire initiators display a hyperbolic energy-power relationship with those limits as the asymptotes, as shown in Figure 4.20.

Generally, where possible, firing systems are designed to provide sufficiently high power so that the initiator will fire near its minimum energy condition. A real example of this relationship is shown in Figure 4.21, the energy-power curve for a Dupont E-78 blasting cap. The points shown are values calculated from the manufacturer's experimental data.

Typical minimum firing energies and bridge resistances for various examples of hot-wire detonators are shown in Table 4.12.

4.4.3 Designs and Considerations for Safety

As can be seen from Table 4.12 in the previous section, firing energies for hot-wire devices are extremely small. The E-78 blasting cap firing energy, which we saw in Figure 4.21, is 0.0076 joules. This is the energy that a 100-watt light bulb uses in just 76 millionths of a second, or put another way, the same as the impact energy on an Oreo® cookie after a fall of 5 in.

Energy can be delivered accidentally to an initiator in two ways: (1) by stray or induced current through the bridge circuit, and (2) by electrostatic or spark discharge through the ignition material from the bridge to the case. Both of these can cause an initiator to function unintentionally (accidentally). These two

Table 4.12 Minimum Firing Energies and Bridge Resistances for Hot-Wire Detonators

Detonator	Bridge Type	Bridge Resistance (W)	Minimum Firing Energy (Millijoules)
T21E1	Carbon	1-10	0.0095
M51	Carbon	1-10	0.028
M48 (APC)	Carbon	1-10	0.042
M48 (CEC)	Carbon	1-10	0.077
T24E1	Tungsten	2-5	0.050
T44/T77	Tungsten	2-500	0.050
T23E1	Tungsten	2-10	0.20
T20E1	Tungsten	2-10	0.37
M36A1	Nichrome	1-10	0.71
MC3089 (match)	?	1	1.5
Typical Blasting Caps	Varies	1.5	5.0
MC2943 (Ignitor)	?	1	10.0
MC3553	?	1	10.0
MC2498	?	1	15.0

modes are independent from each other as far as design considerations are concerned. Both of these modes can be induced by nearby lightning strikes.

Let us consider first the problem of stray or induced current in the bridge circuit. The source could be from meters that produce too much current while measuring the bridge resistance. This problem is taken care of by use of special meters that have current-limiting features, limiting the maximum current to less than 10 milliamperes. A current could be induced in a bridge circuit by radio frequency or microwave transmissions. The electrical leads in the firing circuit can act as an antenna and, if they are inadvertently tuned, can very efficiently convert a radio signal to current. Shipboard radars have been tested with explosive firing systems and found to induce currents as high as 0.5 ampere.

One way to protect the initiator from stray currents is in the initial design of the bridge and header system. Remembering that the firing energy is divided between bridge heating and heat transfer, designs have been made that increase the heat transfer term, thus raising the no-fire current (and hence, minimum power level). This is accomplished by using header materials with greater thermal conductivity as well as increasing the surface-to-volume ratio of the bridge. The latter is accomplished by utilization of ribbon-wire or vapor-deposited metallic film bridges. In this manner, detonators and ignitors have been developed that can withstand continuous bridge currents of 1 ampere without ignition or even degradation of the initiator material. These designs usually utilize 1-ohm bridge resistance, and, therefore, 1 ampere produces 1 watt of power. This type of initiator is called a 1-amp/1-watt "no-fire" hot-wire device. These usually have all-fire current levels at about 3.5 ampere and minimum firing energy in the neighborhood of 50 millijoules.

The 1-amp/1-watt designs, while eliminating much of the problem with stray and induced currents, do not, however, take care of the electrostatic discharge problem. In this case, high electric potential differences between the bridge and the case are the cause of voltage that passes through the ignition material that may initiate the device because the ignition circuitry is not grounded to the case. This is often necessary in many systems due to other constraints. The approach, then, is to make the initiator less prone to internal discharge. This is done in several ways. First, high-resistance leakage paths, external to the ignition material, are provided in some designs. This helps prevent the slow buildup of large potential differences. Next, insulating sleeves are incorporated around the ignition material, forcing any discharge to have to take a much longer path through the initiator. This raises the threshold of the voltage breakdown level. A controlled spark gap can be built around the components' pins that will allow an induced high voltage to arc from pin to case, external to the explosive. And, finally, the ignition material itself can be altered to make it less prone to ignition by arcs. Titanium subhydride fuel in ignition mixes is used for this purpose. Initiators should withstand a 20-kV discharge from a 600-pF capacitor delivered through 500-ohm series resistance, from the shorted pins through the initiator to the case, without initiator function. This condition conservatively simulates the static that could be built up on a human being wearing wool clothing in a dry atmosphere.

4.5 Exploding Bridgewire Detonators

We have looked briefly at initiators that were set into chemical reaction by heating a small volume of the ignition material. The exploding bridgewire (EBW) device, in contrast to the hot-wire initiator, uses shock to initiate reaction. Because of this we do not have to resort to the use of primary explosives to achieve the transition from burning to detonation as in the hot-wire detonators. The shock wave produced by the explosion of the bridgewire initiates detonation directly in a secondary explosive.

If we look at the E_c criteria (Table 4.1) and pop plots (Figure 4.1), we find that only one of the secondaries, PETN, is really easy to initiate at low shock levels, and then only at low densities. Because of this sensitivity to shock (and impact), some people in this field classify PETN with the primaries and not with the secondaries. In any case, PETN is, for right now, the explosive used in EBWs. EBW initiation of other secondaries has been achieved, but the fireset requirements grow so large that PETN is used in most applications.

4.5.1 EBW Detonator Construction

The basic EBW is a very simple device, structurally. The header is shown in Figure 4.22. The lead wires are varied, depending upon the particular design, from deadsoft copper to stiff copper-bronze. The coining in the lead wires in

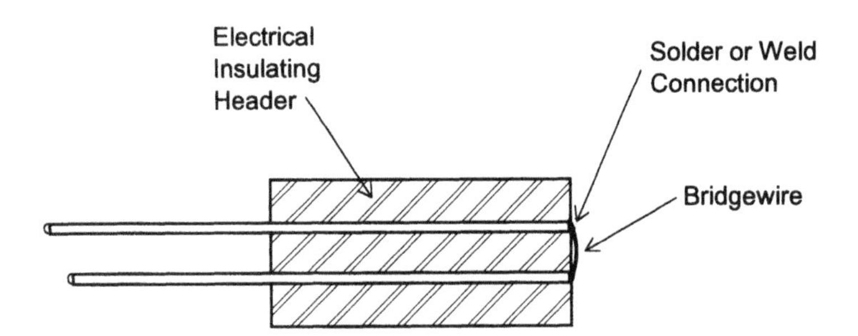

Figure 4.22. A schematic of a typical exploding bridgewire header.

the header is to ensure mechanical strength. The header, itself, is usually molded diallylphthalate, a thermosetting plastic with excellent mechanical and electrical properties. The lead wires are molded directly into the header.

The bridge is usually gold, but could be any number of gold or platinum alloys. The bridge is attached to the lead wires by soldering or welding. The bridge diameter is normally 0.0015 in. (1.5 mils), and the length may vary from as short as 0.01 in. (10 mils) to, in some rare designs, as long as 0.1 in. (100 mils). The most common bridge is 1.5 × 40 mils, gold, soldered; its resistance is approximately 0.01 ohm. A complete detonator is shown in Figure 4.23.

The PETN used is pressed at about 0.9 g/cm^3 and is specially prepared to have an extremely high specific surface area (the higher the specific surface area, the easier it is to initiate). The output pellet is usually tetryl or PBX 9407, prepressed and then inserted into the detonator. A detonator lacking (intentionally) an output pellet is called an initial pressing (IP) detonator.

EBW detonators are available in a range of sizes containing from as little as 10 mg to as much as a gram or more of explosive.

4.5.2 Explosion of the Wire

A quick look at the phenomena that occur in an exploding wire starts with the electrical pulse that supplies the energy. The source of the pulse is the discharge of a capacitor (Figure 4.24). To get the energy into the wire quickly enough to cause the wire to explode, a rise rate of greater than 100 ampere per microsecond is needed. As the current passes through the wire, the resistance rises as the wire heats. Eventually, enough heat is supplied (I^2R heating) to melt the wire. This occurs so fast that the metal has no time to move and is still in the form of a wire, albeit a liquid. Current continues to flow and to heat the melted wire still hotter, until the wire vaporizes. During the melting and vaporization stages, the resistance rises very rapidly, causing a dip in the current. The vaporized wire starts to expand and the current continues to flow, now in the form of an arc. The arc causes ionization of the gases in its path so the resistance drops rapidly;

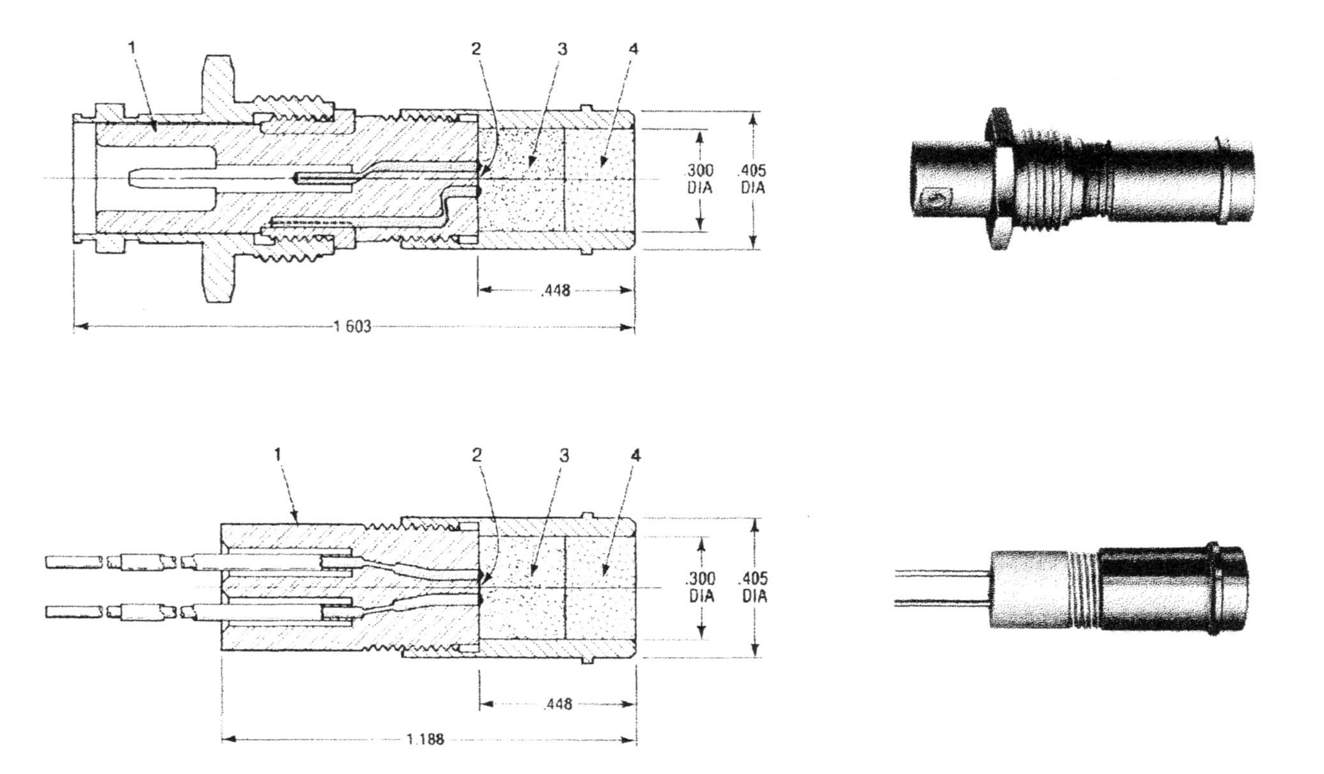

Figure 4.23. Typical EBW detonators, A-connectorized, B-pigtailed (with permission—Reynolds Industries Systems, Inc.). (1) Molded header; (2) bridgewire; (3) low-density PETN; (4) high-density RDX.

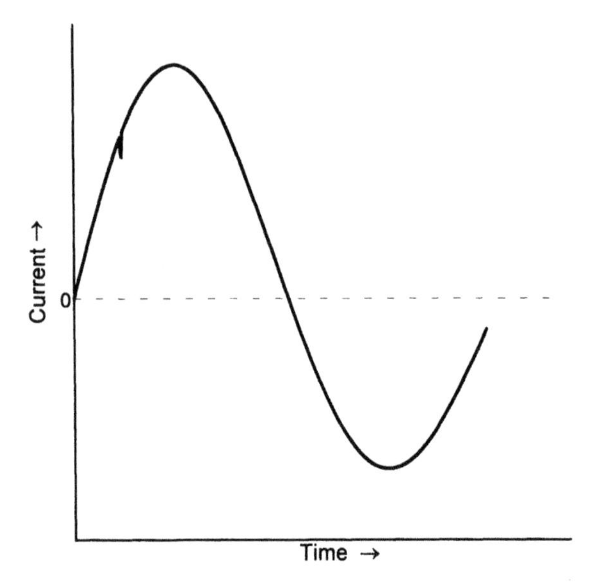

Figure 4.24. Capacitor discharge, the electrical pulse to an exploding wire.

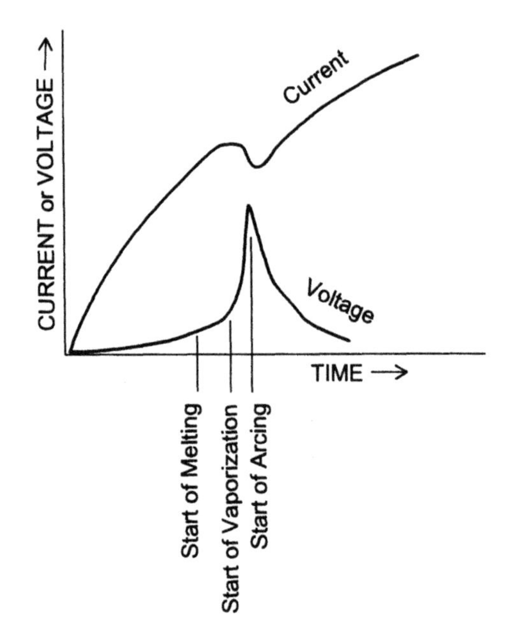

Figure 4.25. Resistance and current versus time in an exploding wire.

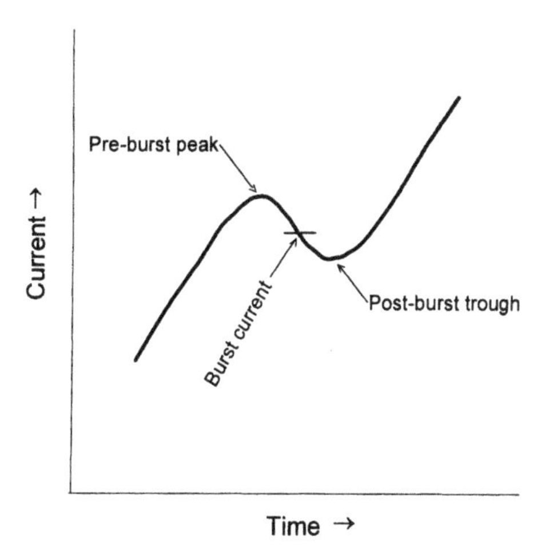

Figure 4.26. Burst current I_b.

the current then rises rapidly. This burst of current is a new burst of heating for the now-expanding vaporized wire and causes the vapors to expand as a shock wave. The resistance and current versus time are shown in Figure 4.25.

Vaporization occurs at the time of maximum resistance, which depresses the current briefly until an arc forms around the wire. The expanding shock wave travels into the explosive. After a short distance, the detonation wave develops (remember the pressure/run distance and the E_c criteria). The first peak on the current wave form is called the preburst peak; this is followed by the dip caused by the resistance spike as the wire bursts as a vapor. At the peak of the resistance spike, the current is falling at maximum rate. This point is called the burst current, I_b, and is shown in Figure 4.26.

The value of the burst current can be approximated by the average of the preburst peak and the burst trough (assuming that the I versus t is symmetric between these points, which is a very good approximation).

For the wire actually to explode and form a shock wave, and for shock to initiate the PETN, the rate of energy input or power must be above some critical value P_{crit}. This, along with properties of both the wire and the explosive, imply a critical current at burst.

For a particular design we can most readily statistically determine a 50% fire condition, or burst current. This is called the threshold burst current, I_{bth}. It has been shown that

$$I_{bth} = Kd/l^{1/2}$$

where d is the diameter of bridgewire (mils), l the length of bridgewire (mils), and K is around 850 A mil$^{-1/2}$. This holds for most bridge materials and PETN

Table 4.13 Available Detonators and Characteristics

Detonator	I_{bth} (amps)	t_t (μs)	Source
RP-1	190	2.95	RISI, Inc.
RP-2	220	1.90	RISI, Inc.
RP-80	180	1.60	RISI, Inc.
RP-83	180	5.15	RISI, Inc.
RP-87	210	1.75	RISI, Inc.
SE-1	180	2.90	DOE
TC217	180	3.00	DOE
TC234	180	4.75	DOE
TC246	180	2.90	DOE

with densities between 0.8 and 1 g/cm^3. Although that is a very limited range for extrapolation, it gives sufficient accuracy for sizing a fireset for a particular detonator, or conversely for choosing a particular detonator for a given fireset. Table 4.13. shows a number of commercially available EBW detonators and their threshold burst current requirements.

4.5.3 Cable and Fireset Considerations

To achieve the very high current and rate of current rise required to exceed the threshold conditions, the impedance of the firing circuitry and cables must be low. Therefore, special low-internal-impedance, high-voltage capacitors are required, and coaxial cables are needed. Normally, cables impose the most severe limitations on the system. The longer the cable, the higher the impedance, and hence the lower the current rise rate. Type "C" cables (coaxial cables rated at 8000+ volts and 30 ohm impedance) can be used up to about 300 feet to initiate EBWs reliably from a 3-kV system using a 1-μF capacitor. For longer cable runs, either higher capacitor voltage or lower cable impedance must be used. The higher the system impedance, the longer the time is from start of pulse until bridge burst.

4.5.4 Safety Aspects of EBWs

If current is supplied to the EBW below threshold, or slower than the critical rise rate, then the bridge will burn out. The bridge may still explode weakly, but it will not produce a high enough shock pressure or E_c to detonate the PETN. In such a case, nothing happens other than that the bridge is now gone. If anything, the detonator is now safer than ever, with no bridge and PETN not being human-body spark sensitive. EBWs are also very safe in the electrostatic discharge mode, easily passing the 20-kV, 600-pF, 500-ohm discharge test.

Drawbacks of EBWs are due primarily to the thermal stability properties of the PETN. This limits the temperature range at which they can be used to the usual "−60°F to +160°F" military requirements.

4.6 Slapper Detonators

The slapper detonator is a fairly recent development. Like the EBW, it depends upon shock and not upon thermal initiation. The shock is from a tiny flyer plate that is driven into the main charge of secondary explosives. The flyer plate can be driven either from a pyrotechnic source or from an electrically exploded foil. The pyro-driven slapper detonator is the older of the two methods, dating back to the late 1950s. Even though that is a long time, very little work was done with the concept, and it has not been used in an off-the-shelf device. In the pyro-driven concept, the flyer was made of aluminum and was about 30 mils (0.030 in.) thick and 100 mils (0.1 in.) in diameter.

In the exploding foil-driven slapper detonators, the flyer is made of Kapton, a thermosetting plastic, and can be as small as 1 to 2 mils thick and 5 mils in diameter. At these sizes, the driver foil can be exploded with the same type pulse as is used for most EBWs, for example, 3 kV from about a 1-μF capacitor. Figure 4.27 shows a small commercial exploding-foil-driven slapper detonator.

When the foil is exploded, it shears out and drives a Kapton disk that accelerates in the barrel to a speed of between 3 and 4 km/s. When the flyer disk impacts the high explosive, it delivers a high-pressure, short-duration shock pulse. The energy fluence developed is in excess of the critical energy fluence

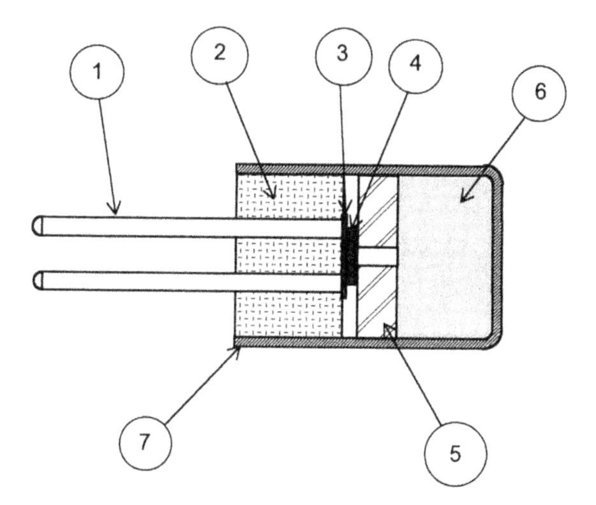

Figure 4.27. Exploding-foil-driven slapper detonator. 1. Lead wires; 2. header; 3. exploding foil; 4. flyer plate; 5. barrel assembly; 6. high-density explosive; 7. cup.

E_c, and the high explosive detonates. Since impact pressures are so high (200 to 300 kbar), the run distance is very short (<0.1 mm).

The advantages of slapper detonators, besides their obviously inherent safety, derive mainly from the fact that the environments in which they are used may be more severe than those in which PETN may be used. Slappers that utilize HNS as the explosive can be hermetically sealed and are able to function reliably at temperatures approaching 300°C at both high pressure and vacuum. One major disadvantage is that because of impedance limits for the explosion of the driving foil, cable lengths are limited to a maximum of about 2 to 3 feet and then only using special flat low-impedance cables.

4.7 Related Reading

1. Bowden, F. P. and Yoffe, A. D. *Initiation and Growth of Explosion in Liquids and Solids*, Cambridge University Press (1952).
2. Bowden, F. P. and Yoffe, A. D. *Fast Reactions in Solids*, Butterworths Scientific Publ. (1958).
3. Chase, W. G. and Moore, H. K. *Exploding Wires*, Plenum Press, New York (1959).
4. Sokolik, A. S. *Self-Ignition, Flame and Detonation in Gases*, Israel Program for Scientific Translations, Jerusalem (1963).
5. *Military Explosives*, TM 9-1300-214/TO 11A-1-34, Departments of the U.S. Army and the U.S. Air Force (Rev. December 1970).
6. *Initiation of Solid Explosives by Impact*, Alfansev, A. and Bobolev, U. translated from Russian by the Israel Program for Scientific Translations (1971).
7. *Explosive Trains*, AMCP706-179 (January 1974).
8. Haase, H. *Electrostatic Hazards*, Verlag Chemie (1977).
9. Catalogs from RISI, Inc., San Ramon, CA.

CHAPTER

5

Scaling in Design and Analysis

In this chapter we will examine the use of scaling in various forms that help in the design or analysis of explosive systems. We will start with geometric similitude scale modeling, a tool used for developing experiments and tests of small models from which we can derive design information for larger systems. Then we will look at mathematical scaling based on pre-existing databases that allow us to calculate designs or derive data for systems of any size. These will include the acceleration of metal by explosives, the prediction of shock pressures in both air and water from explosive charges, the prediction of the size of craters formed in hard soil from surface blasts, and finally the performance of conical-shaped charges.

5.1 Geometric Similarity

Scaling by geometric similarity is a technique whereby the behavior of a given system can be determined by conducting experiments on a smaller model of that system. The smaller system is called the model, and its dimensions and properties are noted by the subscript "0." The larger system is called the prototype, and its dimensions and properties are noted by the subscript "1." The model and the prototype must be exactly similar in geometric design. Every dimension in the model is scaled down by the same factor, "S." Thus, if we scaled a box with a scaling factor $s = 4$, then the prototype would be four times the height, four times the width, four times the depth, and four times the wall thickness of the model. If the model is to be tested to destruction, let us say we are looking for the internal pressure that will burst open the seams, the nails or screws must be geometrically scaled also by a factor of four.

Geometric similarity also demands that material properties be exactly the same in both the model and prototype, if those properties influence the behavior that we are studying, for example, tensile strength or density or whatever.

Let us look at some examples and see how this works. First, consider a very simple system such as a cube, as shown in Figure 5.1.

The scaling factor is S. Therefore, $X_1 = SX_0$. How does the surface area of the prototype scale with that of the model? Surface area of a cube $= 6X^2$

$$A_0 = 6X_0^2$$

$$A_1 = 6X_1^2$$

$$X_1 = SX_0$$

$$X_1^2 = S^2X_0^2$$

$$\frac{A_1}{A_0} = \frac{6X_1^2}{6X_0^2} = \frac{6S^2X_0^2}{6X_0^2} = S^2$$

The areas in the prototype are scaled as the square of the scaling factor. How about volumes? Volume V of a cube $= X^3$.

$$V_0 = X_0^3$$

$$V_1 = X_1^3$$

$$X_1 = SX_0$$

$$X_1^3 = S^3X_0^3$$

$$\frac{V_1}{V_0} = \frac{X_1^3}{X_0^3} = \frac{S^3X_0^3}{X_0^3} = S^3$$

Volumes scale as the cube of the scaling factor. Simple. How about mass? Mass equals density times volume, $m = \rho V$.

$$m_0 = \rho V_0$$

$$m_1 = \rho V_1$$

$$\frac{m_1}{m_0} = \frac{\rho V_1}{\rho V_0} = \frac{\rho S^3 V_0}{\rho V_0} = S^3$$

The mass scales as the cube of the scaling factor. This is pretty obvious; let us move on. Suppose our system is a spherical pressure vessel, as shown in Figure 5.2 We will burn a propellant inside the vessel and determine the final pressure. In both the model and prototype, we use the same kind of propellant.

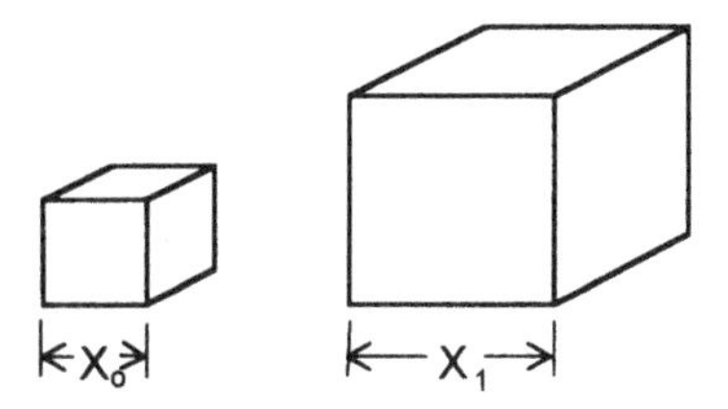

Figure 5.1. Model and prototype cubes.

The propellant charge is also geometrically similarly scaled.

$$r_1 = Sr_0$$

$$r_{c1} = Sr_{c0}$$

r_c is the radius of the charge and r is the radius of the vessel. This allows us to find the final pressure in each vessel. Propellant properties pertinent to the problem are density ρ_c, gas evolution V_g (cm^3/g), and isochoric flame temperature T_v. From the above, we know the volumes of the vessels V_0 and $V_1 = S^3 V_0$. We also know the masses of the propellants m_0 and $m_1 = S^3 m_0$. The volume of gas evolved from the propellant is $V_{\text{STP}} = V_g m$ (at STP), and the final temperature of the gas will be T_V. The initial temperature is T_{STP}, and the initial pressure is P_{STP}. Let us now use the ideal gas equation, $PV = nRT$.

$$P_0 = \frac{V_g m_0 T_V P_{STP}}{V_0 T_{STP}}$$

$$P_1 = \frac{V_g m_1 T_V P_{STP}}{V_1 T_{STP}}$$

$$\frac{P_1}{P_0} = \frac{(V_g m_1 T_V P_{STP})/(V_1 T_{STP})}{V_g m_0 T_V P_{STP})/(V_0 T_{STP})} = \frac{m_1 V_0}{m_0 V_1}$$

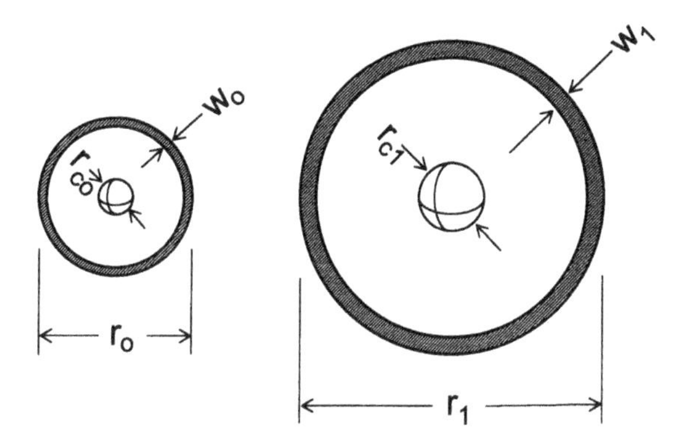

Figure 5.2. Model and prototype of a spherical pressure vessel containing a propellant charge.

We saw above that $m_1 = S^3 m_0$.

$$V_1 = S^3 V_0$$

So,

$$\frac{P_1}{P_0} = \frac{S^3 m_0 V_0}{m_0 S^3 V_0} = 1$$

We see that pressures are equal in both the model and prototype in geometric similarity modeling. Going a bit further, let us look at the stresses in the walls of these scaled vessels. The wall thickness was also scaled; so these are w_0 and $w_1 = Sw_0$. The spherical hoop stress in a thin-walled ball is

$$\sigma = \frac{Pr}{2w}$$

where P is the internal pressure, r the mean radius, and w the wall thickness. So,

$$\sigma_0 = \frac{P_0 r_0}{2w_0} \quad \text{and} \quad \sigma_1 = \frac{P_1 r_1}{2w_1}$$

We saw above that $P_0 = P_1$ and were given $r_1 = Sr_0$ and $w_1 = Sw_0$, so

$$\frac{\sigma_1}{\sigma_0} = \frac{(P_1 r_1)/(2w_1)}{(P_0 r_0)/(2w_0)} = \frac{(P_0 S r_0)/(2Sw_0)}{(P_0 r_0)/(2w_0)} = 1$$

So now we see that stresses are equal in both model and prototype.

If both of these vessels were made of the same metal at the same heat treatment, they would both burst at the same pressure. Complex vessels that cannot be designed analytically can be modeled and tested in this manner very economically. Let us look at another dimension, time. Suppose our system is now a gun firing over a short range into a target, as shown in Figure 5.3.

The velocity of the bullet, U, is the same in both model and prototype. The time of travel from the muzzle to the target is $t = X/U$, so

$$t_0 = X_0/U \quad \text{and} \quad t_1 = X_1/U$$

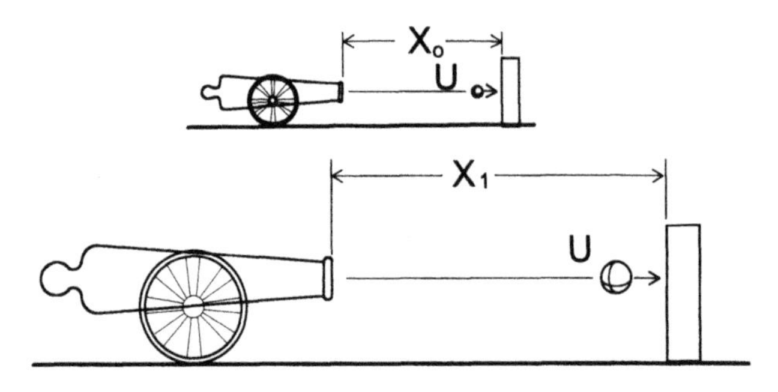

Figure 5.3. Model and prototype of an idealized gun.

and of course the scaling factor is $X_1 = SX_0$, so

$$\frac{t_1}{t_0} = \frac{X_1/U}{X_0/U} = \frac{SX_0/U}{X_0/U} = S$$

So time scales with the scaling factor. Without belaboring the point with more complicated equations, it can be shown that for scaled guns with scaled propellants as well as scaled bullets, the muzzle velocities would be equal. Geometrically similar scaled bullets with equal mass densities have the same drag coefficients, and it will be shown later that the entire trajectory is scaled such that velocity loss is equal over scaled distances.

This technique has been successfully used to test many ordnance and explosive systems economically such as target damage in concrete walls from artillery projectiles (in that case, the aggregate in the concrete was also scaled so that the targets would be geometrically similar). The key word above is economics. Small-scale testing is much less expensive than large-scale testing. Consider the cost of materials alone. A 5:1 scale model weighs $\frac{1}{125}$ as much as its prototype, and materials are purchased by weight. A 10:1 scale model weighs $\frac{1}{1000}$ the weight of the prototype.

Scaling by geometric similarity does have some limits, however. There are some processes that do not scale. For instance, spalling of metal plates that were in contact with a detonating explosive is a process that is not conducive to scaling. The type of spall and its distance from the metal's first free surface are functions, among other things, of the slope of the Taylor wave behind the detonation front. This slope does scale somewhat for thin explosive charges (thin in the direction of detonation) but eventually becomes constant at a thickness greater than a few inches. So the spalling scales up to that thickness but not beyond. The explosive in that case is not adhering to our stated rule of properties' similarity. Another case of this type is scaling a system containing a cylindrical explosive charge. As we saw earlier, the detonation velocity and hence the C-J pressure decrease with decreasing diameter. Thus for geometries where the diameters of the model are small, the explosive properties are changing with the scaling. The model may even scale down to below the critical diameter. This is also true of systems that use thin sheets of explosive. As long as one is aware of such limitations, excellent results are obtained with the scaling technique.

5.2 Accelerating Metal with Explosives

This method is used to predict the velocity to which explosives can accelerate materials placed in contact with them. R. W. Gurney originally proposed this simple technique in predicting the expansion rate of cylindrical bomb casings during detonation. J. E. Kennedy, while at Sandia National Laboratories, extended the technique by applying it to additional geometries. The only properties needed for prediction of velocity are: (1) the Gurney velocity, $(2E)^{1/2}$ (explosive property), (2) the weight of the metal piece to be moved, M (mechanical property), (3) the weight of the explosive, C (mechanical property), and (4) the weight of a tamper, N (mechanical property).

The following equations yield the dimensionless velocity, $v/(2E)^{1/2}$, as a function of the load factor, M/C, for various geometries. The geometries, which have been defined by Kennedy, are subdivided into two categories: symmetric and asymmetric. Symmetric geometries are defined such that the explosive and metal flyer are symmetric about a centerline or plane. These geometries and the associated equations are shown in Figures 5.4 and 5.5. In Figure 5.6, these relationships are evaluated numerically and plotted as dimensionless velocity versus metal-to-charge mass ratio.

A list of Gurney velocities is shown in Table 5.1. These values are based mainly on cylinder-test data. For other explosives, or these same explosives

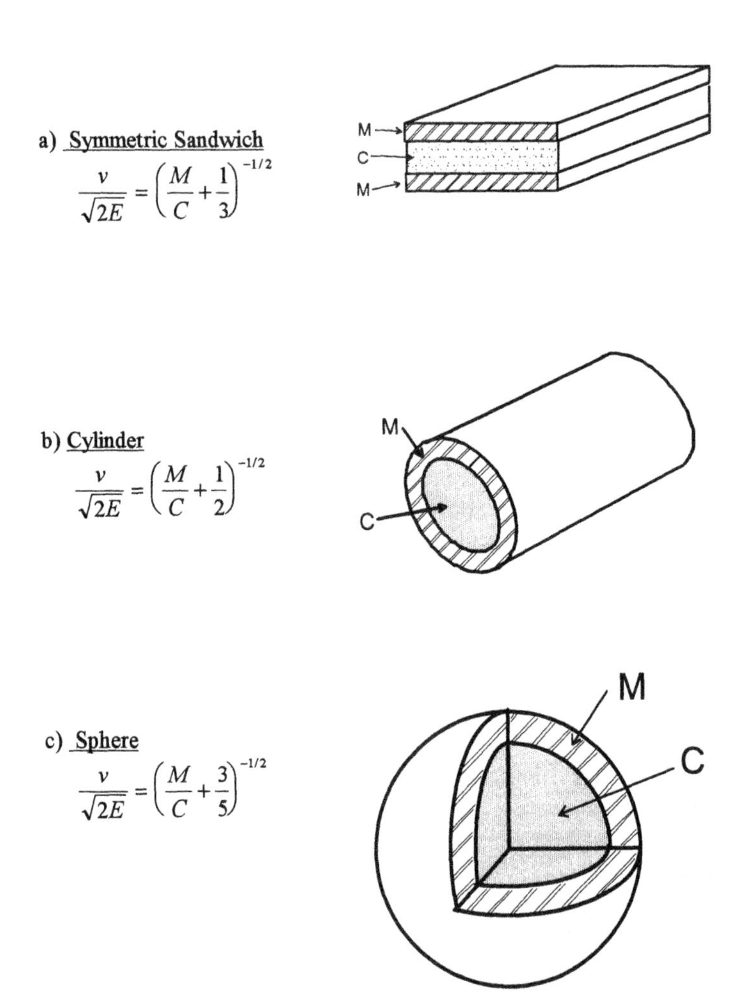

Figure 5.4. Symmetric geometries.

a) Open Face Sandwich

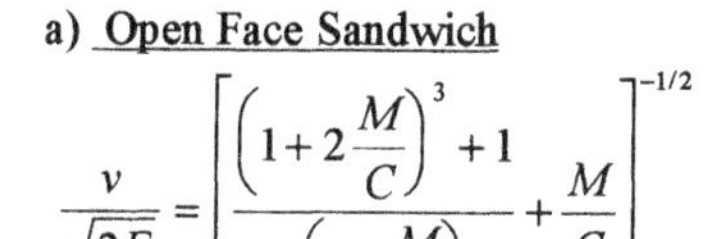

$$\frac{v}{\sqrt{2E}} = \left[\frac{\left(1 + 2\frac{M}{C}\right)^3 + 1}{6\left(1 + \frac{M}{C}\right)} + \frac{M}{C} \right]^{-1/2}$$

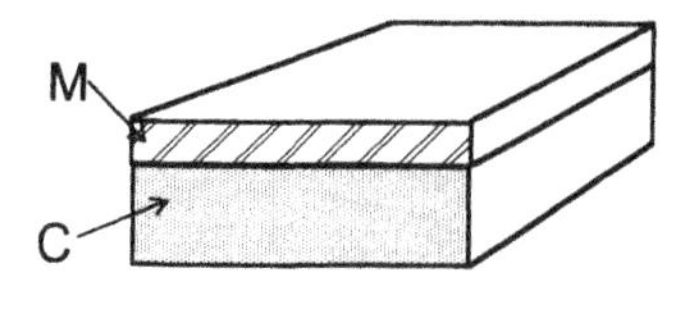

b) Asymmetric Sandwich

$$\frac{v}{\sqrt{2E}} = \left[\frac{1 + A^3}{3(1 + A)} + \frac{N}{C} A^2 + \frac{M}{C} \right]^{-1/2}$$

where:

$$A = \frac{1 + 2\frac{M}{C}}{1 + 2\frac{N}{C}}$$

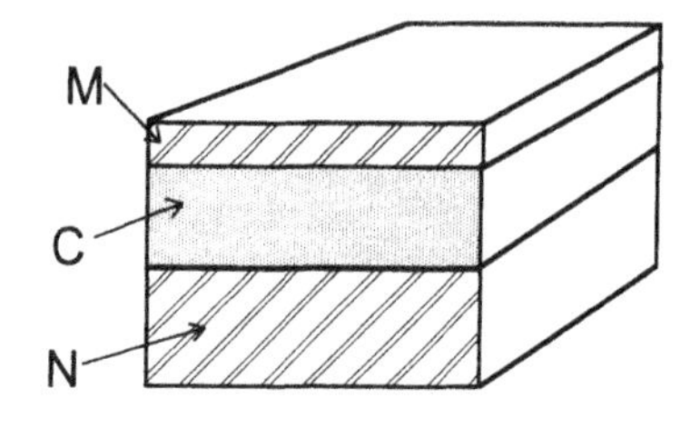

Figure 5.5 Asymmetric geometries.

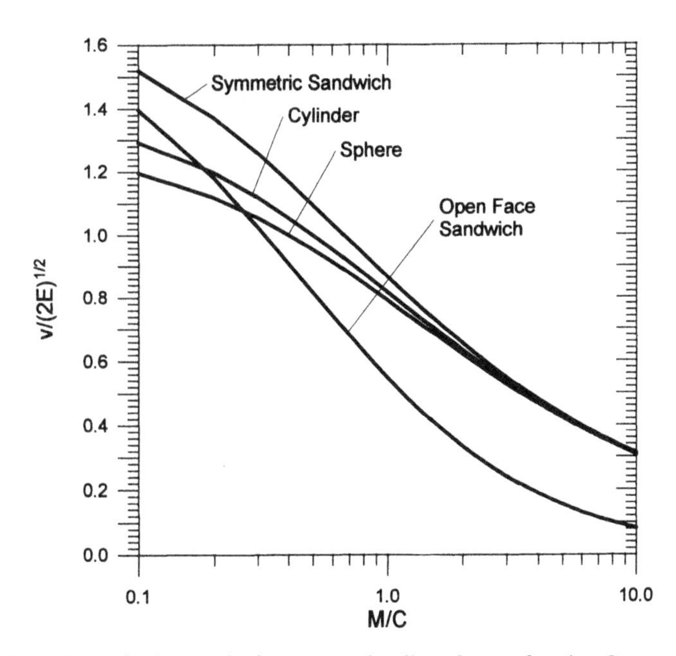

Figure 5.6. Dimensionless velocity versus loading factor for the Gurney-Kennedy method.

Table 5.1 Gurney Velocities for Various Explosives

Explosive	Density (g/cm³)	$(2E)^{1/2}$ (km/s)
Composition B-3	1.71	2.70
HMX	1.89	2.97
PBX 9404	1.84	2.90
PETN	1.76	2.93
RDX	1.77	2.93
TACOT	1.61	2.12
Tetryl	1.62	2.50
TNT	1.63	2.40

at other densities, a good estimate of Gurney velocity can be obtained from $(2E)^{1/2} = D/3$, where D is the detonation velocity of the explosive. This estimate is usually within 7% of that observed in experiments.

Example 5.1

What will the expansion velocity be of a steel cylinder, 2 in. inside diameter, and wall thickness of 0.2 in., filled with Composition C-4 explosive. The density of the steel is 7.85 g/cm³, and the density of the Composition C-4 is 1.59 g/cm³.

Solution The masses of the steel cylinder and explosive filler are equal to their volumes times their respective densities.

$$M = L\left(\frac{\pi}{4}\right)(d_0^2 - d_i^2)\rho_{stl}$$

$$C = L\left(\frac{\pi}{4}\right)d_i^2\rho_{exp}$$

$$\frac{M}{C} = \frac{\rho_{stl}(d_0^2 - d_i^2)}{\rho_{exp}d_i^2} = \frac{7.85(2.4^2 - 2.0^2)}{1.59(2.0^2)} = 2.17$$

From Table 3.5 in Chapter 3, we see the detonation velocity of Composition C-4 at 1.59 g/cm³ density is 8.04 km/s.

$$(2E)^{1/2} = 8.04/3 = 2.68 \text{ km/s}$$

For a cylinder geometry

$$v = (2E)^{1/2}\left(\frac{M}{C} + \frac{1}{2}\right)^{-1/2} = 2.68(2.17 + 0.5)^{-1/2} = 1.64 \text{ km/s (5380 ft/s).}$$

5.3 Shock Waves in Air

When an explosive charge detonates in air, the expanding detonation product gases push on the surrounding air, forcing out a shock wave. The larger the explosive charge, the higher the peak shock pressure at any given distance from

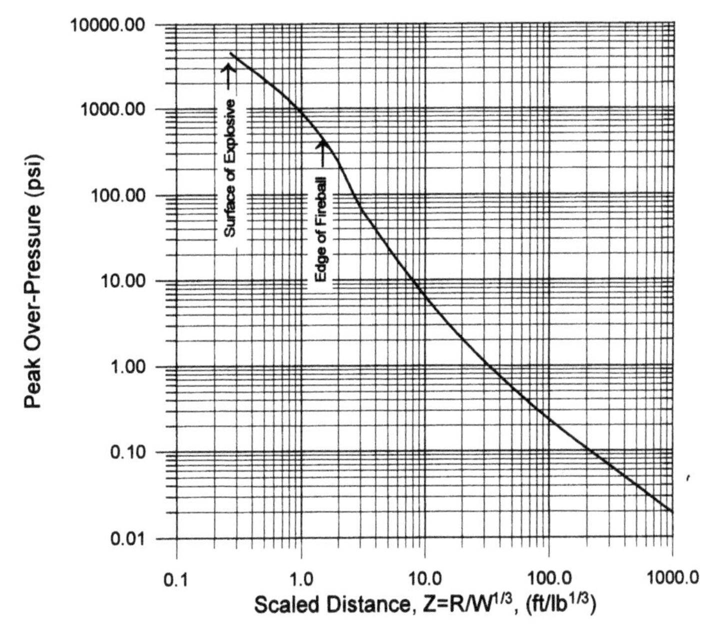

Figure 5.7. Peak shock pressure versus scaled distance for TNT blasts at sea level.

the charge. The peak pressure at any distance for any size charge of any explosive can be quite accurately estimated based upon scaling experiments using TNT. In Figure 5.7, we see the peak blast or shock pressure P plotted versus the scaled distance, $R/W^{1/3}$. In this relationship, P is in psi, R, the distance from the center of the charge, is in feet, and W, the weight of the charge, is in pounds of TNT. If we are interested in finding the peak pressure at some given distance from a different explosive, then the TNT equivalent weight of that explosive is used as W. The *TNT* equivalent weight can be found in Table 3.8 or can be estimated by any of several methods. One method is to use the ratio of the heat of detonation of the subject explosive to that of *TNT*. Another is to estimate by use of the *C-J* properties.

$$W(TNT \cdot equivalent) = W_{\text{exp}}\left(\frac{P_{CJ}}{\rho_0}\right)_{\text{exp}} \Big/ \left(\frac{P_{CJ}}{\rho_0}\right)_{TNT}$$

The scaling curve in Figure 5.7 was based on experiments using *TNT* at 1.64 g/cm^3 density. The P_{CJ} for *TNT* at that density is about 21.0 GPa. Therefore,

$$W(TNT \cdot equivalent) = 0.078 W_{\text{exp}}\left(\frac{P_{CJ}}{\rho_0}\right)_{\text{exp}}$$

where W is in pounds, P_{CJ} is in GPA, and ρ_0 is in g/cm^3.

It should also be noted that this scaling curve is for air blast at sea level. At higher altitudes, the shock pressure at any given distance from a given charge would be somewhat lower; the higher the altitude, the lower the pressure.

Example 5.2

We detonate a charge in the air of Composition C-4, which weighs 40 pounds. What is the resulting shock pressure at 5 feet, 10 feet, 100 feet?

Solution First we must find the TNT equivalent weight of the charge. From Table 3.5 we see that the density of Composition C-4 is 1.59 g/cm^3 and the *CJ* pressure is 25.7 GPa. So the *TNT* equivalent weight is

$$W_{TNT} = 0.078(40 \text{ pounds})\left(\frac{25.7 \text{ GPa}}{1.59 \text{ g/cm}^3}\right) = 50.4 \text{ pounds}$$

The scaled distance at 5 feet is $Z = (5 \text{ ft})/(50.4 \text{ pounds})^{1/3} = 1.35$. At that distance the peak pressure (from Figure 5.7) is found to be 500 psi. At $Z = (10 \text{ feet})/(50.4 \text{ pounds})^{1/3} = 2.7$, $P = 100$ psi; and at $Z = (100 \text{ feet})/(50.4)^{1/3} = 27$, $P = 1.3$ psi.

5.4 Shock Waves in Water

As with shock waves in air, when an explosive charge is detonated under water, it drives a shock wave into the surrounding water. Also like shock waves in air, scaling experiments were conducted to develop a scaling base. It was found by R. H. Cole in the 1940s that the following empirical equation predicted the peak shock pressure at any distance under water:

$$P = K\left(\frac{W^{1/3}}{R - R_0}\right)^{\alpha}$$

where P is the peak shock pressure in psi; W is the weight of explosive in pounds; R_0 is the radius of the charge in inches (this assumes a spherical charge); R is the distance from the center of the charge in inches (note that in the air blast scaling section before this that R was in feet; in the Cole equation above it is in inches!), K is a constant, different for different explosives, and α is a constant that is also different for different explosives but averages about 1.14 (ranges from 1.12 to 1.16) for most of them. Values for K can be found in the open literature for only a few explosives; however, it can be closely estimated by $K = 1.5 \times 10^5(1 + \Delta H^0_{\text{exp}})$, where ΔH^0_{exp} is the heat of explosion of the explosive being used in kcal/g.

These estimating equations are accurate at most distances; however, they cannot be used for values of R less than $2R_0$. That close to the charge, the scaling equation grossly overestimates the pressure.

Example 5.3

What is the peak shock pressure at 30 feet from the detonation of 25 pounds of pentolite?

Solution From Table 3.2, we find that for pentolite, $\rho_0 = 1.70$ g/cm^3 (0.0614 pounds/in.^{3}) and $\Delta H^0_{\text{exp}} = 1.40$ kcal/g. First, we will have to assume that the charge is spherical, and then we can find R_0. For a sphere, vol $= \frac{4}{3}\pi R^3$, and we can find the volume from the weight and density,

$$\text{vol} = W/\rho_0 = (25 \text{ pounds})/(0.0614 \text{ pounds/in.}^3) = 407 \text{ in.}^3$$

$$R = \left(\frac{3V}{4\rho}\right)^{1/3} = \left(\frac{3(407)}{4\pi}\right)^{1/3} = 4.6 \text{ in.}$$

Next we will find the value of K:

$$K = 1.5 \times 10^5(1 + \Delta H^{\circ}_{\text{exp}}) = (1.5 \times 10^5)(1 + 1.4) = 3.6 \times 10^5.$$

Now we can find P:

$$P = K\left(\frac{W^{1/3}}{R - R_0}\right)^{\alpha} = (3.6 \times 10^5)\left(\frac{25^{1/3}}{12 \times 30 - 4.6}\right)^{1.14} = 1513 \text{ psi}$$

5.5 Craters from Explosives

When an explosive charge is detonated in contact with the ground, it forms a crater in the ground. The larger the charge, the larger the crater. The crater is formed by the shock wave, coupled to the ground, breaking up the ground material. The combination of the surface relief wave behind the shock and rapid expansion of the detonation product gases heave the shattered earth up into the air. Some of this dirt falls back around the crater, forming a lip, and some falls back down into the crater, partially refilling it. Most craters are from three to five times wider than they are deep, the depth being variable partially because of the uncertainty of the amount of ejecta fallback. The radius of the crater is quite predictable if one knows the soil/rock type and properties as well as the explosive properties, shape or geometry of the charge, and the height of burst above the ground or depth of burial below the ground. Quantifying all of these parameters is beyond the scope of this text. However, we will examine the most common case, which is for the surface burst of a spherical charge in hard, undisturbed, relatively dry dirt.

By surface burst, we mean that the sphere of explosive is placed so that its center is at ground level, half the charge above the ground, and half below. This fixes position, ground type, and charge geometry, leaving only the explosive weight and type as variables. Numerous experiments have led to the following scaling relationship between the weight of explosive and the apparent crater radius R_a. The apparent radius is the radius at the original ground level and is smaller than the lip radius.

$$R_a = kW^{1/3}$$

where R_a is the apparent crater radius in feet, W is the explosive charge weight in pounds, and k is a constant, different for each explosive.

It has been found by these authors that $k = 0.46 + 0.027\ P_{CJ}$, where P_{CJ} is the C-J pressure of the explosive in GPa. Combining these gives us a general expression for surface burst craters in hard dirt.

$$R_a = (0.46 + 0.027 P_{CJ})W^{1/3}$$

If the explosive were fired above the ground, the crater would be smaller and smaller the higher the height of burst. When the center of the charge is only three diameters above the ground level, the crater dimensions shrink to only 5% of those for a surface burst.

For buried charges, the crater dimensions increase as the depth of burial increases up to a given scaled depth. Beyond that point the weight of the overburden becomes too great to be thrown away, and the crater size then goes down with continued increasing depth of burial until a depth is encountered where no crater forms and the explosion is completely contained. The optimum depth of burial (DOB), that is, that which gives maximum crater dimensions, is about DOB = $1.5W^{1/3}$, (where DOB is in feet and W is in pounds). The approximate depth for total containment is DOB = $6W^{1/3}$.

Example 5.4

Suppose we have a sphere of PBX-9404 that weighs 50 pounds. What would the apparent crater radius be if we detonated it as a surface burst? How deep would we have to bury it to completely contain it?

Solution Once again looking at Table 3.2 for explosive properties, we find the C-J pressure of PBX-9404 is 37.5 GPa. Then

$$R_a = (0.46 + 0.027 \times 37.5)(50)^{1/3} = 5.4 \text{ feet}$$

In order to contain this charge completely, we would have to bury it

$$DOB = 6W^{1/3} = 6(50)^{1/3} = 22 \text{ feet}$$

5.6 Conical-Shaped Charges

5.6.1 Configuration

The shaped charge is generally a conical shape (Figure 5.8). The liner material is usually copper, aluminum, or mild steel, although glass is also sometimes used. The explosive is usually pressed or cast.

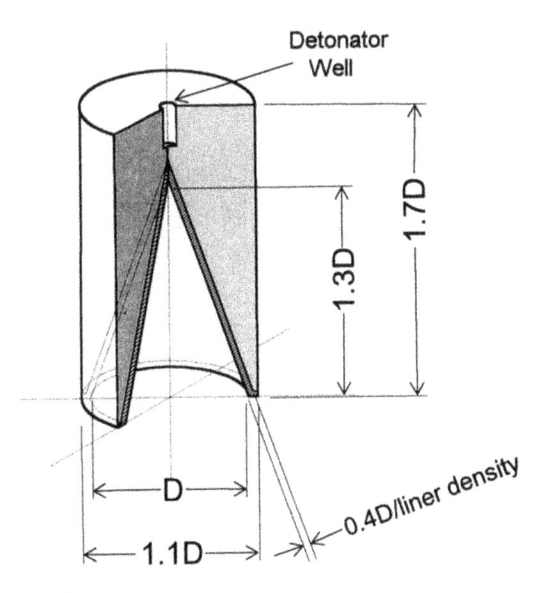

Figure 5.8. Conical-shaped charge.

The charge works by explosively collapsing the liner, which forms a high-velocity jet of liner material. The formation of the jet is rather complex to model mathematically, so let us look at the phenomena qualitatively.

5.6.2 Jet Formation

As the explosive detonation wave passes over the liner, the liner is accelerated at some small angle to the explosive liner interface (Figure 5.9).

Nearer the apex of the cone, the M/C ratio is lower, the liner velocity is higher. When the liner material converges at the center line (or axis) of the charge, the surface material is squeezed out at high velocity (Figure 5.10).

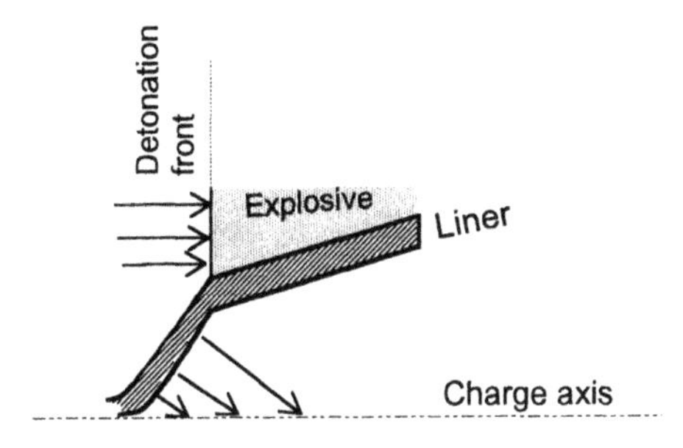

Figure 5.9. Acceleration of the liner during the passage of an explosive detonation wave.

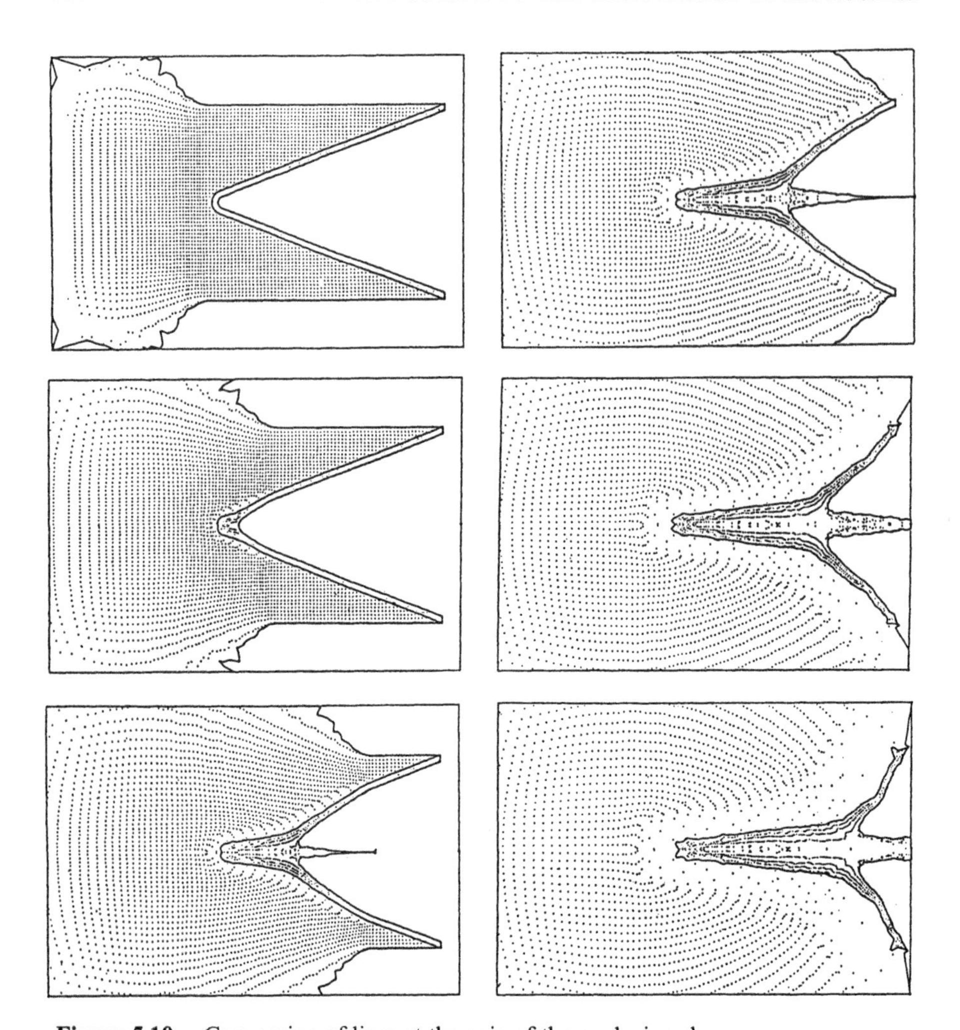

Figure 5.10. Converging of liner at the axis of the explosive charge.

This "squeezed-out" material forms the jet. Since the material closest to the apex was at higher velocity, the portion of the jet that comes from that area is also highest in velocity. Therefore, we have a jet with a velocity gradient where the leading tip is moving faster than the rear. This gradient is assumed to be linear. The remaining material, which is the bulk of the liner, forms a heavy "slug" that follows the jet at much lower velocity (Figure 5.11).

Since there is a velocity gradient along the jet, the farther it travels, the longer it gets. Many of the particles that form the jet have slightly different directions of flight because of minor inhomogeneities in the charge and liner. After travel of several diameters (of the original charge), the jet begins noticeably to break up.

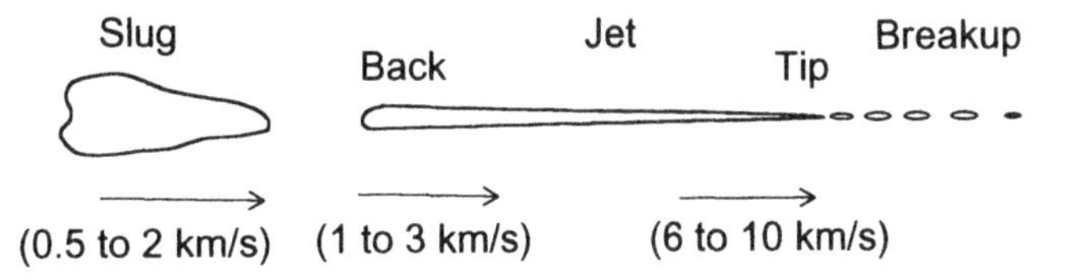

Figure 5.11. Jet configuration.

5.6.3 Effect on Target

The erosion of a target by a penetrating jet is very similar to what one sees when a stream of water from a hose is squirted into a bank of dirt. Material dislodged at the deepest part of the hole turns to mud and flows back along the walls of the hole (Figure 5.12).

Metal targets under shaped-charge jet attack behave like fluids because at the impact velocities of the jet, both jet and target at the interface are at several megabars pressure, well into the plastic region for almost all materials. This erosion process continues until the entire jet has been used up or until the target has been perforated.

5.6.4 Effect of Standoff

By "standoff" we mean the distance of the base of the charge from the target. This is usually expressed in charge diameters (Figure 5.13). At very short standoffs, the jet is still very short; it has not had time to form, or "stretch"; therefore penetration into the target is less than optimal.

At very long standoffs, the jet is breaking up, and each particle is hitting farther and farther off center and is not contributing to the penetration at the center of the target (Figure 5.14). Figure 5.15 depicts a typical penetration versus standoff relationship for a conical-shaped charge.

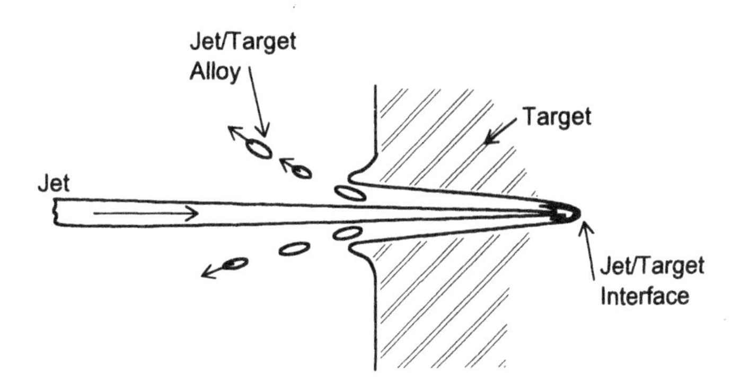

Figure 5.12. Erosion of jet through target.

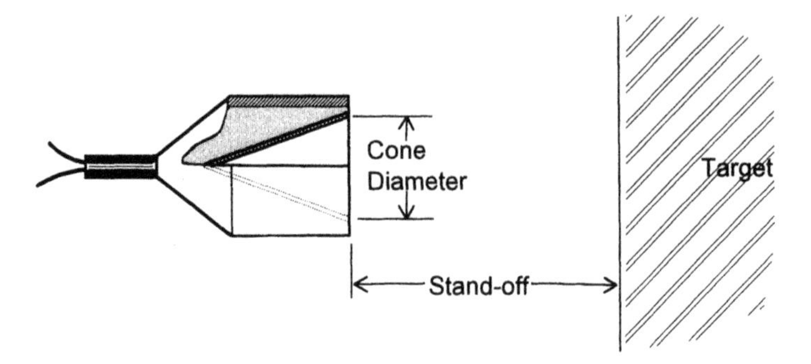

Figure 5.13. Standoff from the target.

5.6.5 General Observations

1. Greatest penetration is obtained with charges that have cone angles of around 42°.
2. Optimum standoff is usually between 2 and 6 charge diameters.
3. Penetration is normally around 4 to 6 charge diameters and could go as high as 11 or 12.
4. Optimum liner thickness is about three percent of the cone base diameter for soft copper. This can be scaled for changes in density (change of material) by keeping weight constant. That is, lower density liners should be thicker.

5.6.6 Penetration Model

We looked briefly at what shaped charges are and how they work qualitatively. Now let us consider a simple model that will help quantify some of these observations. The model assumes that both the jet and the target are ideal liquids (that

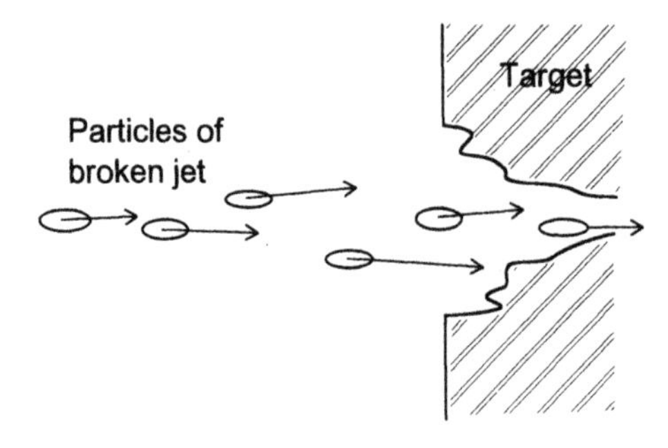

Figure 5.14. Example of longer than optimal standoff.

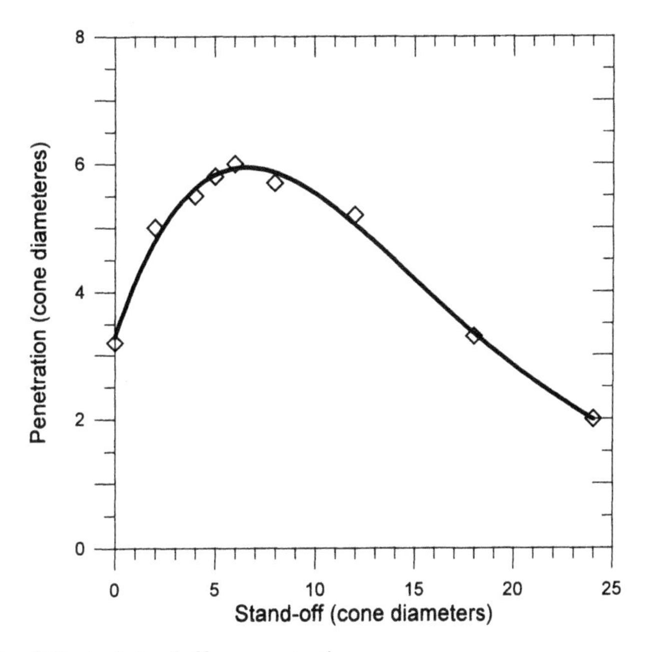

Figure 5.15. Effect of standoff on penetration.

is, they do not exhibit any viscosity). This is not a bad assumption, because most metals are far into the plastic region and do indeed behave like liquids at the impact pressure at the interface of jet and target (several megabars). The next assumption is that the jet is traveling at a constant, uniform velocity. We know this is not true; but it is surprising how well the model holds even with this oversimplifying assumption. The last assumption is that the jet, a liquid, is in the form of a rod. Figure 5.16 illustrates a jet penetrating a target. It does not show the jet or target material flowing backward and out of the hole, just the progress of the jet and the hole.

From the Bernoulli theory, the pressure of the jet at the jet/target interface is

$$P = \frac{1}{2} \rho_j V_R^2 \tag{5.1}$$

where V_R is the relative velocity of the jet and the end of hole (the end of the hole is receding from the jet at the penetration velocity V_p); that is,

$$V_R = V_j - V_P$$

Equation (5.1) then becomes:

$$P = \frac{1}{2} \rho_j (V_j - V_P)^2 \tag{5.2}$$

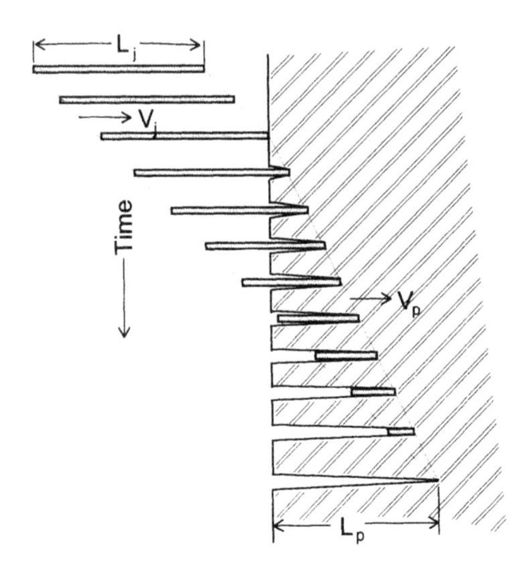

Figure 5.16. Jet penetration of a target.

The pressure in the target at this same interface is, of course, the same, and is

$$P = \frac{1}{2}\rho_T V_P^2 \tag{5.3}$$

The time it takes to complete the penetration is found from the penetration velocity and the depth of penetration.

$$t = \frac{L_P}{V_P} \tag{5.4}$$

This time is the same as that required to "use up" the jet, or the time it takes for the back of the jet (see Figure 5.16) to reach the interface (the end or bottom of the hole).

$$t = \frac{L_j + L_P}{V_j} \tag{5.5}$$

Let us combine Eqs. (5.2) and (5.3):

$$\rho_j(V_j - V_P)^2 = \rho_T V_P$$

$$\frac{\rho_j}{\rho_T} = \left(\frac{V_P}{V_j - V_P}\right)^2 \tag{5.6}$$

$$\left(\frac{\rho_j}{\rho_T}\right)^{\frac{1}{2}} = \frac{V_P}{V_j - V_P}$$

Now combine Eqs. (5.4) and (5.5):

$$\frac{L_P}{V_P} = \frac{L_j + L_P}{V_j}$$

$$\frac{L_P}{V_P} - \frac{L_P}{V_j} = \frac{L_j}{V_j}$$

$$\frac{L_P(V_j - V_P)}{V_P V_j} = \frac{L_j}{V_j} \tag{5.7}$$

$$L_P(V_j - V_P) = L_j \frac{V_P V_j}{V_j} = L_j V_P$$

$$L_P = L_j\left(\frac{V_P}{V_j - V_P}\right)$$

Now, combining Eqs. (5.6) and (5.7), we get

$$L_P = L_j\left(\frac{\rho_j}{\rho_T}\right)^{1/2} \tag{5.8}$$

This equation is the idealized penetration equation, which allows us to predict penetration performance in various targets and helps analyze shaped-charge designs. Notice that the equation implies that the depth of penetration L_p is independent of the jet velocity, and depends only on the length of the jet and the relative density of jet to target. Experiments show that this is true and holds fairly well for jet velocities above 3 km/s.

Example 5.5

A conical-shaped charge with a copper liner is fired into a steel target at the optimum standoff for that charge. The jet penetrates 4.5 in. into the target.

a. What do you estimate the jet length would be, at optimum standoff?
b. What penetration would this shaped charge achieve at the same standoff, but into an aluminum target?

Solution

a. We can get the jet length by rearranging Eq. (5.8)

$$L_j = \frac{L_P}{(\rho_j/\rho_T)^{1/2}} = \frac{4.5}{(8.93/7.89)^{1/2}} = 4.2 \text{ in.}$$

Now solve for the penetration using the above jet length, the density of the jet, and the density of the aluminum target in the penetration equation, $L_P = L_j(\rho_j/\rho_T)^{1/2} =$

$4.2(8.93/2.78)^{1/2} = 7.5$ in. We could have found the penetration into the aluminum also by forming the ratio of

$$\frac{L_{\text{P,Alumn}}}{L_{\text{P,Steel}}} = \frac{L_j(\rho_j/\rho_{\text{Alumn}})^{1/2}}{L_j(\rho_j/\rho_{\text{Steel}})^{1/2}},$$

giving us

$$L_{\text{P,Alumn}} = L_{\text{P,Steel}}\left(\frac{\rho_{\text{Steel}}}{\rho_{\text{Alumn}}}\right)^{1/2} = 4.5\left(\frac{7.89}{2.78}\right)^{1/2} = 7.5 \text{ in.}$$

5.6.7 Shelf Hardware

Many shaped charges, both commercial and military, are available "off-the-shelf." Some are far more efficient than others in relation to depth of penetration as a function of size or explosive weight. This spread of penetration efficiency is due, not to poor design, but because each charge was designed for a particular application, and not all were optimized with penetration alone in mind. Figure 5.17 presents the performance of a compendium of over a hundred different-shaped charges, showing the ranges one should expect of penetration versus weight of charge based upon what is available off the shelf.

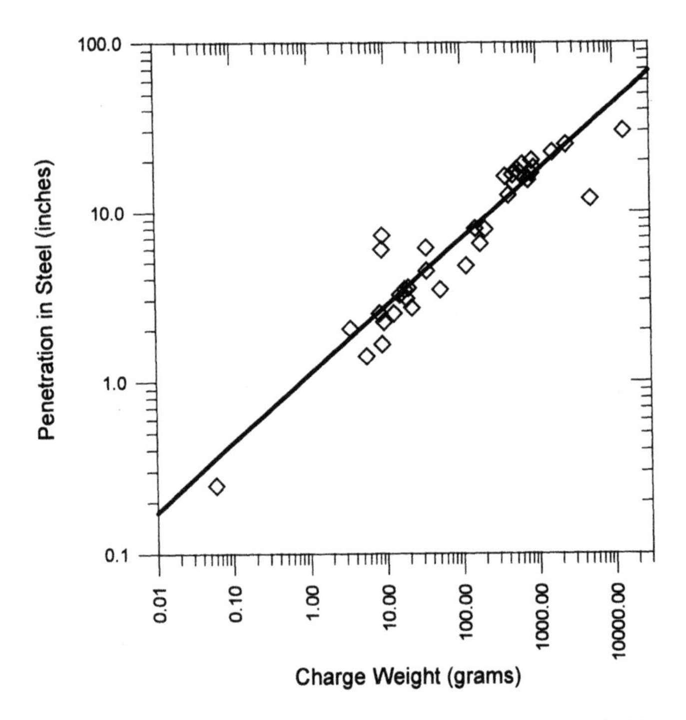

Figure 5.17. Penetration in steel versus charge weight for a number of different-shaped charges.

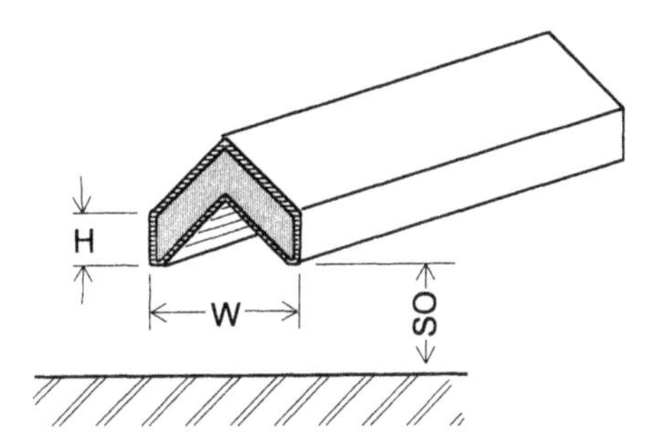

Figure 5.18. Linear-shaped charge.

5.6.8 Linear-Shaped Charges

The linear-shaped charge differs from the conical in that the jet is formed from the collision of two plates rather than from the collapse of a cone. This is achieved by constructing the charge in a chevron cross-sectional shape with a long axial dimension perpendicular to that. This is shown in Figure 5.18.

The jet formed from the linear-shaped charge is like a long knife blade. It behaves the same as the jet from a conical-shaped charge except that it cuts a long groove or channel in a target rather than a conical hole. Most linear-shaped charges are manufactured by impact or linear extrusion. Typical sheath (and hence liner) materials are lead, copper, or aluminum; explosive fillers are typically RDX or HNS. Approximate loading and performance are (referring to Figure 5.18):

$W \approx C^{1/2}/32$ (W in in., C is the core loading in grains per foot)
$H \approx W/4$
SO (optimum) $\approx 0.2W$ (for lead sheathed)
$0.7W$ (for copper sheathed)
$1.0\ W$ (for aluminum sheathed)
Penetration into steel $\approx 0.6W$ (for lead sheathed)
$0.7W$ (for aluminum sheathed)
$0.9W$ (for copper sheathed).

Linear-shaped charges are used extensively in demolition work for cutting plates and girders, and assemblies where water is excluded from the jet-forming area are used in maritime salvage operations for cutting ships into manageable pieces.

5.7 Related Reading

1. Langhaar, H. L., *Dimensional Analysis and Theory of Models*, John Wiley & Sons Inc. (1951).

2. Rinehart, J. S. and Pearson, J., *Explosive Working of Metals*, Macmillan Press (1963).

3. Sedgwick, R. T., Gittings, M. L., and Walsh, J. M., ''Numerical Techniques for Shaped Charge Design,'' in *Behavior and Utilization of Explosives in Engineering Design*, 12th Annual Symposium, jointly by NM Section ASME and UNM College of Engineering (March 1972).

4. Crossland, B., *Explosive Welding of Metals and its Application*, Clarendon Press, Oxford (1982).

5. Baker, W. E., *Explosions in Air*, Wilfred Baker Engineering Publications, San Antonio (1983).

6. Blazynski, T. Z., *Explosive Welding, Forming and Compaction*, Applied Science Publishers, London (1983).

7. Walters, W. P. and Zukas, J. A., *Fundamentals of Shaped Charges*, John Wiley & Sons, Inc. (1989).

8. Baker, W. E., Westine, P. S. and Dodge, F. P., *Similarity Methods in Engineering Dynamics*, Elsevier Press, Amsterdam (1991).

CHAPTER

6

Off-the-Shelf Explosive Devices

In this chapter we will look at some off-the-shelf explosive devices starting with linear explosive cords and charges, and then examining a number of explosives/mechanical devices.

6.1 Linear Explosive Products

The linear explosive products are manufactured in continuous lengths. They are divided here into three categories: (1) deflagrating cords, (2) detonating cords, and (3) linear-shaped charges.

6.1.1 Deflagrating Cords

These products all burn rather than detonate. They are used for delays, ignition sources, and deflagrating distribution systems.

6.1.1.1 Safety Fuse

This is a fiber-jacketed cord impregnated with asphalt with an orange plastic coating for water proofing (Figure 6.1). The core is black powder diluted with nonexplosive materials to control the burning rate. It is approximately 0.25-in. in diameter and fits snugly into the open end of nonelectric blasting caps. It is used extensively in the mining industry. It can be ignited with a common match

Figure 6.1. Safety fuse deflagrating cord.

or any convenient flame source such as a pyrotechnic pull ignitor or spitter fuse. Safety fuse comes in two propagation rates: 120 and 90 seconds per meter.

6.1.1.2 Igniter Cord

This material (Figure 6.2) has, in addition to a thin fiber wrap, a wrap of fine woven steel wire and is brown in color. It is made up in $\frac{1}{8}$ in. diameter and burns in two speeds: Type A burns at 30 seconds per meter and Type B at 60 seconds per meter. Unlike safety fuse, igniter cord spits flame and sparks out sideways as it burns. As its name implies, it is used to ignite other materials or other cords.

There is a different burning cord that looks similar to ignitor cord, called quarry cord. That cord burns more rapidly, about 3 seconds per meter. Quarry cord is no longer being manufactured.

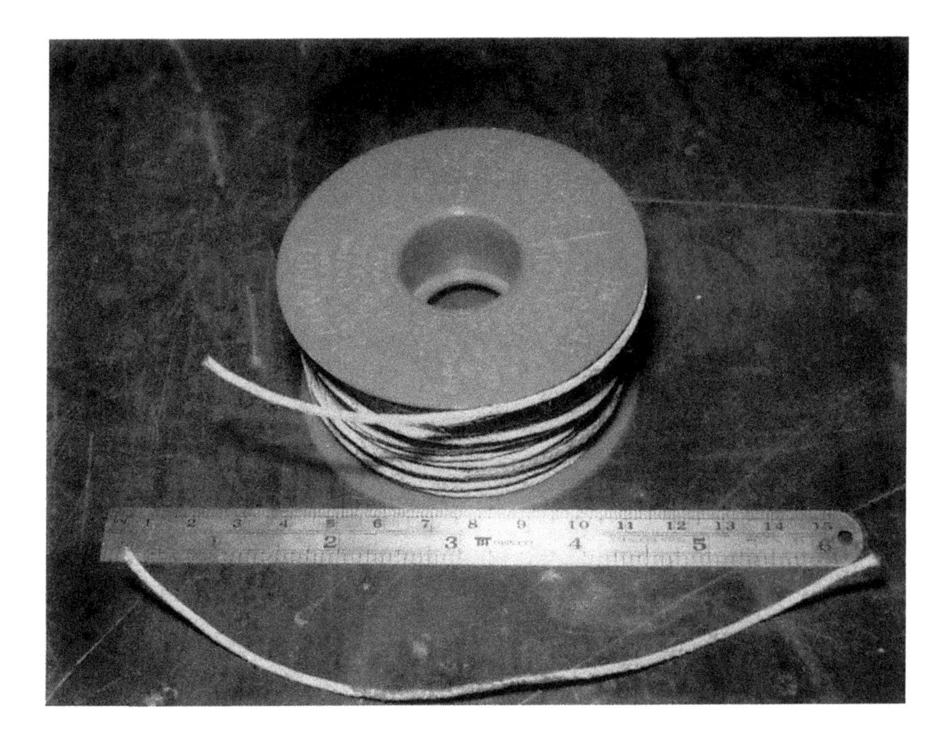

Figure 6.2. Igniter cord.

6.1.1.3 Pyrofuze®

Also called Exo-Wire, this material does not really burn. The fuse is made of a wire of aluminum with a jacket of palladium (Figure 6.3). It is made by extrusion through wire-drawing dies, starting with a billet of palladium pipe with a rod of aluminum inside it. It is ignited by heating the fuse to the melting point of aluminum. The aluminum melts and the palladium dissolves in it in an exothermic alloying reaction, forming a palladium-aluminum alloy.

This alloying reaction gives off heat, about 325 calories per gram, which makes the reaction propagate. While reacting, the material glows white and shoots out glowing sparks. It is usually supplied as a braid having from 5 to 9 strands of 1- to 3-mil-diameter wire. The propagation rate is approximately 3 seconds per meter. This fuse is used in some special delay lines, and because of its very photogenic sparking, as a special-effects pyrotechnic in motion pictures.

6.1.2 Detonating Cords

These materials all detonate. They are used mainly as detonation transfer lines, or submillisecond delays.

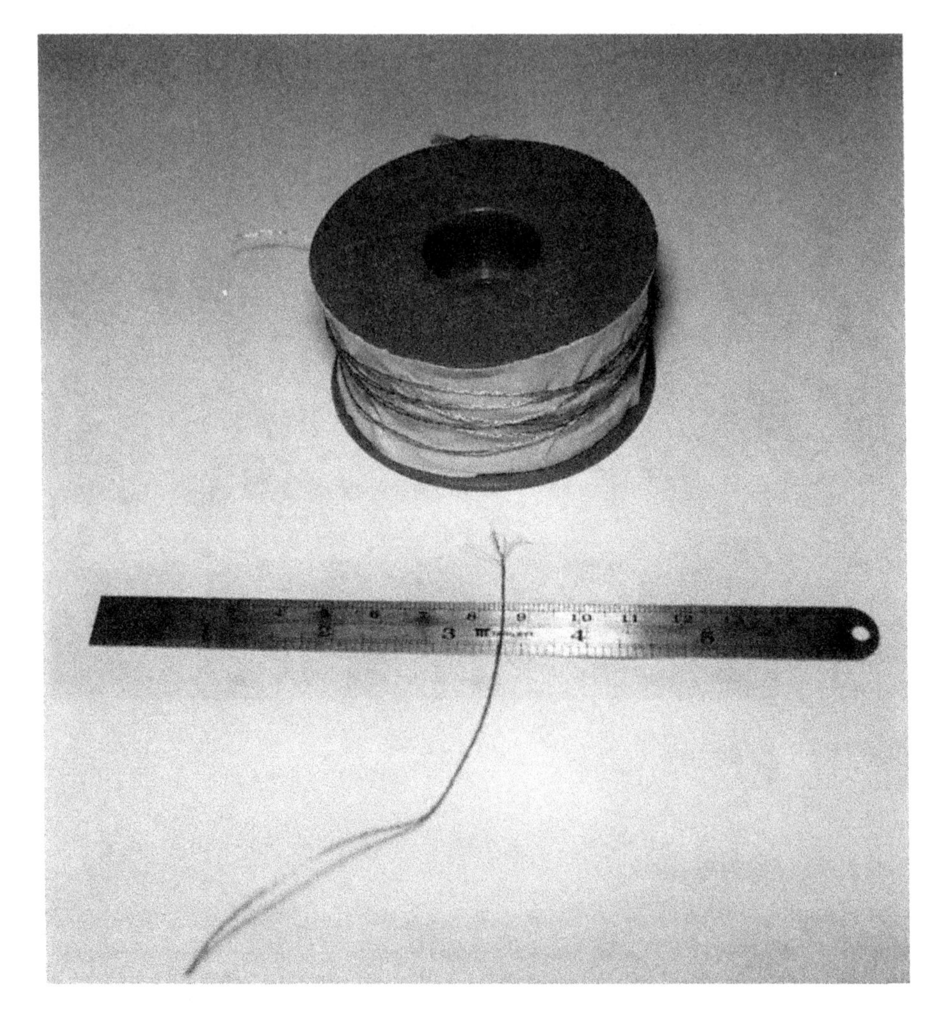

Figure 6.3. Braided Pyrofuze® wire.

6.1.2.1 Primacord

This is probably the most familiar as well as the highest production volume item among the detonating cords (Figure 6.4). It is made by weaving a cloth sock around a core of PETN (or sometimes RDX). The sock is then covered with a plastic sheath to provide waterproofing and additional strength. Both PETN and RDX are white-colored powders, the RDX is often dyed pink in these cords in order to make identification easier. Primacord is supplied in loading weights of from 7.5 grains per foot (~1.6 grams per meter) up to 400 grains per foot (~85 grams per meter). Detonation velocities vary from 7.0 to 7.5 kilometers per

Figure 6.4. Primacord.

second (23,000–25,000 feet/second). Primacord is used for detonation distribution in mining and quarrying. It can be used to detonate most dynamites directly.

6.1.2.2 MDF

This product derives its name from "mild detonating fuse" (Figure 6.5), mild, because it comes in core loading as low as 0.5 grains per foot (0.1 grams per meter). The sheath materials are usually lead, but versions come with aluminum or silver sheaths also. The core explosives are usually PETN, but versions are available with lead azide, RDX, HMX, HNS, and HNAB. Detonation velocities vary between 6.5 and 8.0 km/s, depending upon the core explosive, the core density, and the sheath wall thickness.

MDF is used in detonation distribution systems and manifolds in complex weapons. In addition, a version that has a heavy yarn sheath around the metal, which makes it self-contained, is used in detonation transfer and manifolds in

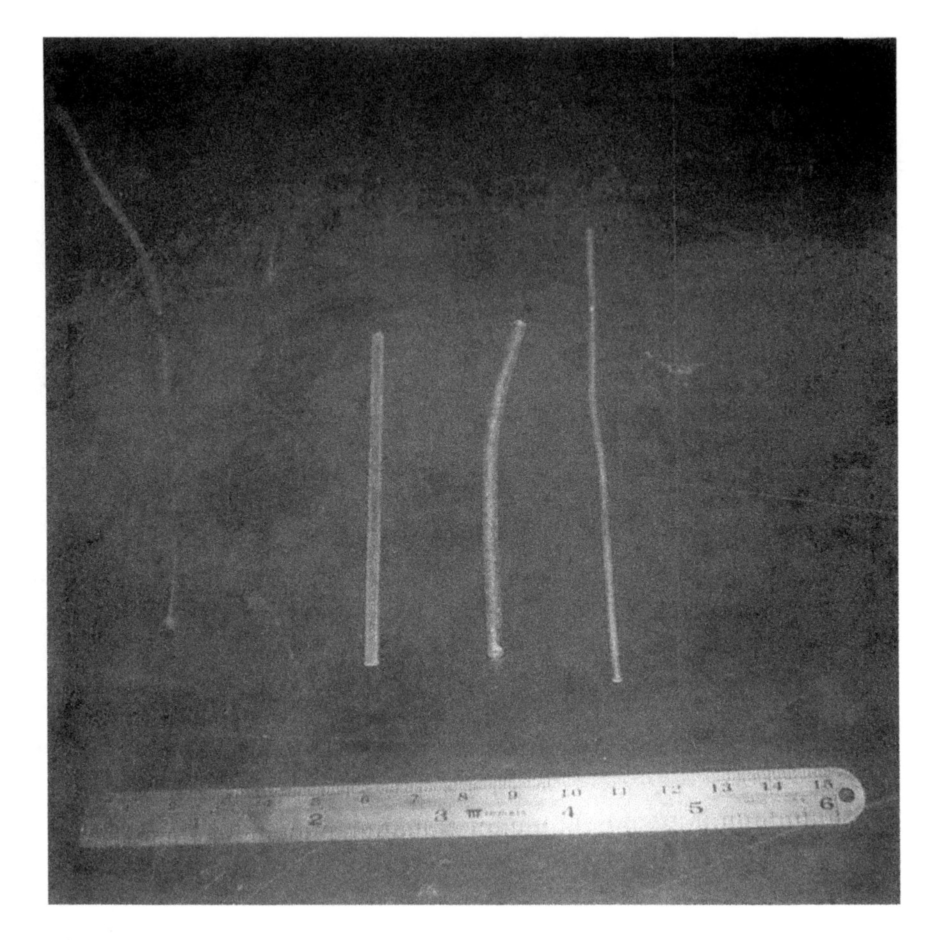

Figure 6.5. Mild detonating fuse.

aircraft escape systems. MDF is manufactured by extruding an explosive-filled metal pipe through sets of wire-drawing dies.

6.1.2.3 Detacord

This is not a very common cord. It is made by extruding detasheet through small-hole dies. It is, in all properties, the same as Detasheet.

6.1.2.4 Shock Tube

Shock tube, or Nonel®, is a fairly new product on the market (Figure 6.6). Patents for Nonel® are owned by Nobel-Abel in Sweden and licensed to several com-

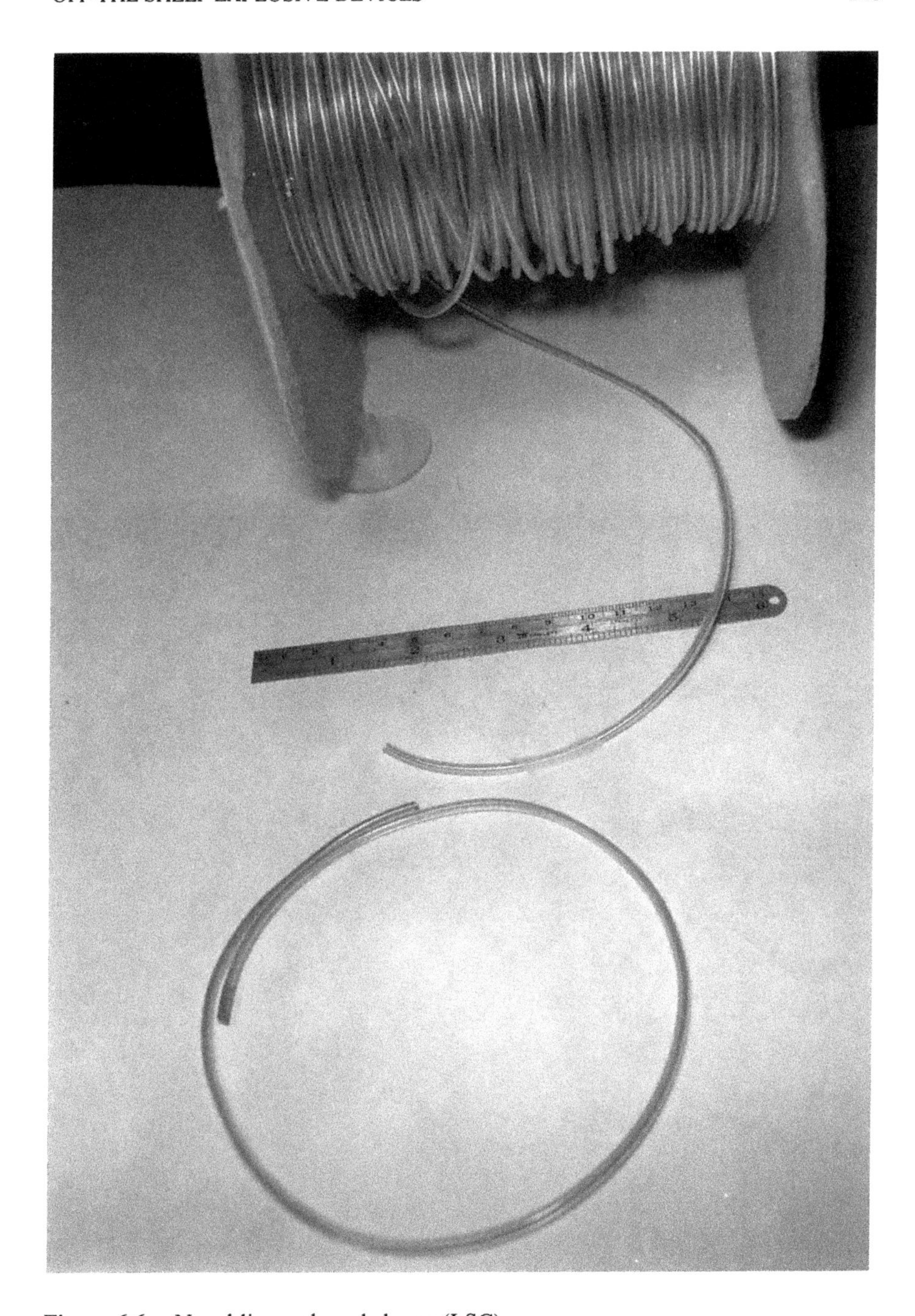

Figure 6.6. Nonel linear-shaped charge (LSC).

panies in the United States. It is a small-diameter, hollow plastic tube, coated on the inside with a very thin layer of explosive. The explosive is a mixture of extremely fine-particle-size HMX and aluminum. Loading is approximately 0.2 grains per foot (0.042 grams per meter, or 0.15 pounds per mile). Because the explosive is so thin, the reaction has large side losses and therefore detonates quite slowly, at approximately 2 km/s (around 6.5 feet per millisecond). The tube completely contains the detonation; it can be held in one's hand while detonating without any danger of injury. Shock tube is used in place of either Primacord or safety fuse, wherever the location dictates a need for its special characteristics. When used to initiate other explosives, it must have a nonelectric blasting cap attached to its output end.

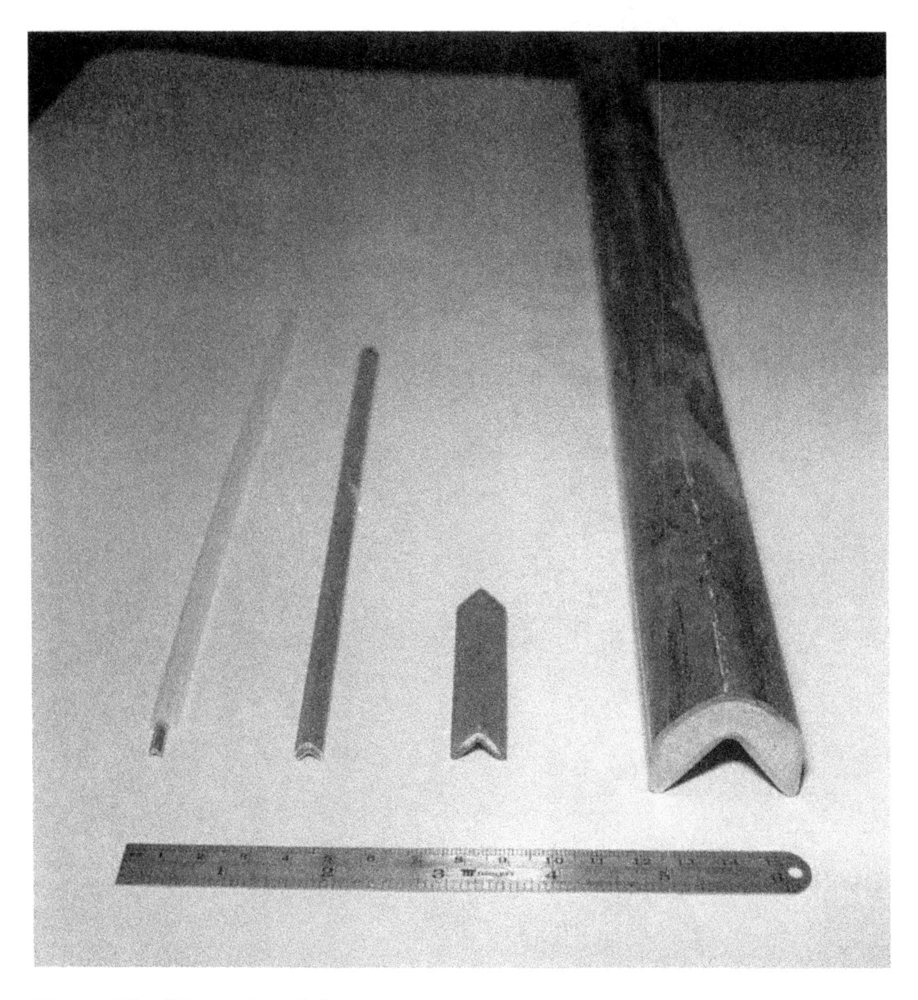

Figure 6.7. Linear-shaped charge.

6.1.2.5 Linear-Shaped Charge

The linear product shown in Figure 6.7 is similar to MDF, except instead of having a round cross section, it has a chevron-shaped cross section.

Like MDF, this charge has a metal sheath of lead, aluminum, or copper. The smaller sizes are drawn through dies; the larger sizes are swaged. The core can be any one of several explosives, including PETN, RDX, HMX, or HNS. In the larger sizes where the linear-shaped charge (LSC) is swagged to shape, the explosive is usually waxed RDX. The nomenclature generally used depends upon the metal sheath; thus:

lead sheathed = FLSC
aluminum sheathed = ALSC
copper sheathed = CLSC

Some manufacturers use the names ''jet-cord'' and ''x-cord'' for LSCs. Like conical-shaped charges, a jet is formed from the ''V'' portion of the charge during detonation. The same general behavior as with conical-shaped charges is observed; that is, there is an optimum standoff, and the cutting or penetration is a function of both target and jet densities. The jet velocities are much lower in LSCs, on the order to 2 to 3 km/s. The optimum standoffs are very short as compared to conical charges. The depth of penetration is also small relative to conical-shaped charges. Cutting performance of various LSCs are listed in manufacturers' catalogs, and it is best to use those data when they are available. However, when such data are not readily at hand, dimensions and cutting ability can be estimated by using the relationships given in Section 5.6.8.

Example 6.1

You do not have a manufacturer's catalog handy, but you want to estimate the charge weight, or size, or an aluminum-sheathed linear-shaped charge (ALSC) to cut a 0.5-in.-thick steel plate. You know from experience with this particular steel that if you penetrate the plate about halfway, it will crack the rest of the way through by itself.

Solution All you need is a charge to penetrate 0.25 in. From Section 5.6.8, we have for ALSC: relative penetration in steel (in charge widths) = 0.68, and relative weight factor (charge width)/(core load)$^{1/2}$ = 0.0312 (in.)(gr/ft)$^{1/2}$

Penetration = 0.68 $\times$ W; therefore, W = Penetration/0.68 = 0.37. W/C$^{1/2}$ = 0.0312; therefore, C = (W/0.0312)^2 = 141.

Now, LSCs do not come in such odd sizes; so you would pick the next available size increment up from that. That would probably be 150 grains per foot. (When this was checked against one of the manufacturer's catalogs, we found that 150 grains/foot RDX-ALSC penetrates 0.245 in. into steel. That is not bad estimating.)

Some manufacturers' catalogs refer to ''severance'' instead of penetration, the total cutting ability of an LSC from penetration plus fracture due to the shock in the target material. Severance can be as high as twice the penetration for some target materials.

6.2 Mechanical/Explosive Devices

There are a number of explosive items that do not look, upon a cursory inspection, as if they have anything at all to do with explosives. Most of these devices are self-contained and pose no injury hazard if they function prematurely. Some of these items, however, do detonate violently.

6.2.1 Exploding Bolts

One of the major mechanical/explosive devices that is truly dangerous upon functioning is the exploding bolt. These bolts are used in systems where a one-time, nearly instantaneous detachment is required. Each bolt has a detonating charge inside which, when initiated, causes the bolt to rupture at a specific plane by radial tensile fracture. Some very small steel fragments are formed from around the fracture plane, but the major hazard would be from the two main halves being thrown apart by the detonation. The bolts are available in a range of sizes; Figure 6.8 shows three of them.

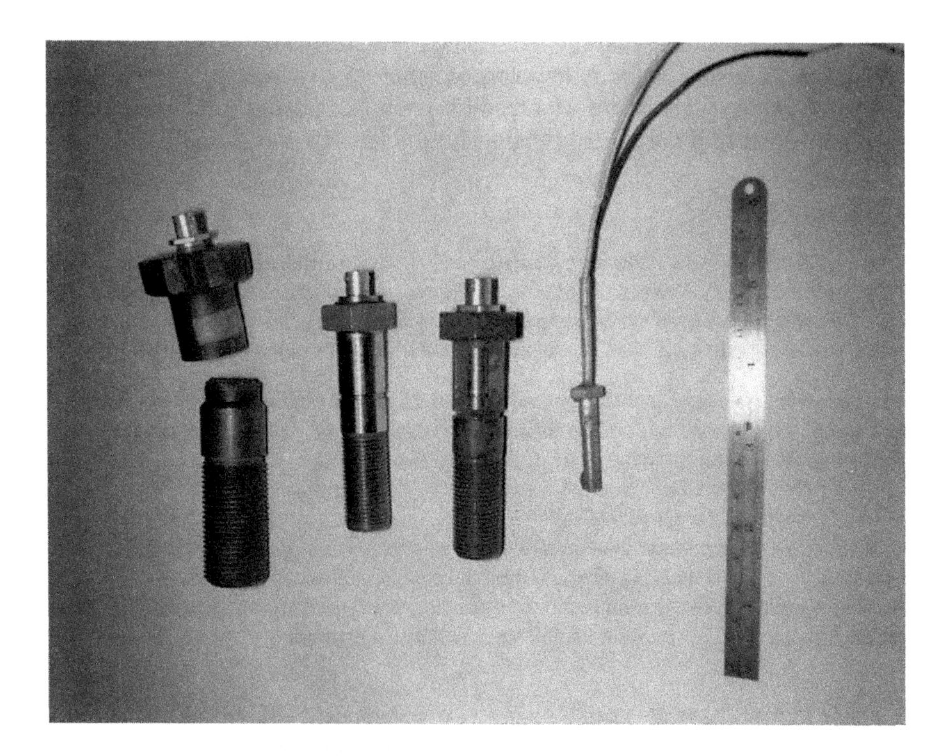

Figure 6.8. Exploding bolts of different sizes.

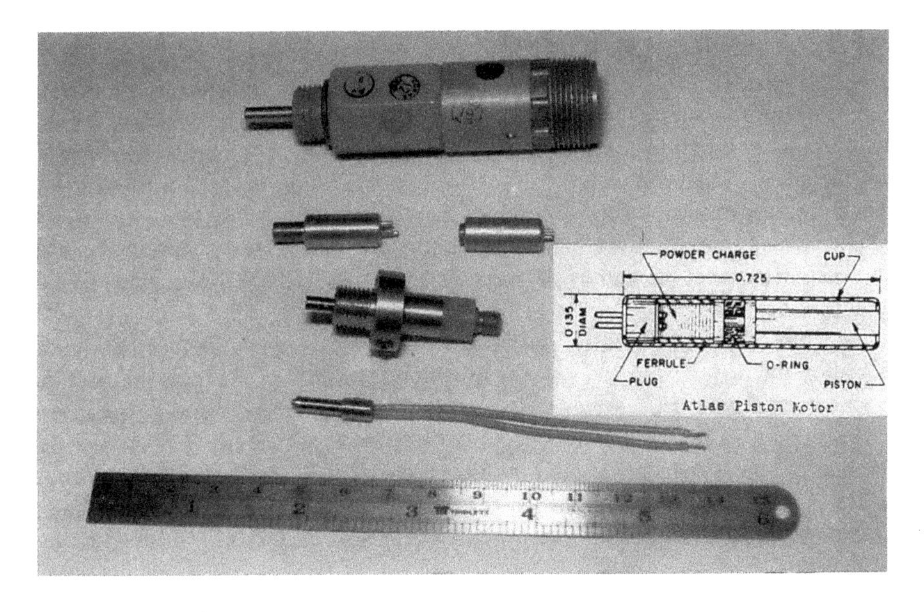

Figure 6.9. Piston motors.

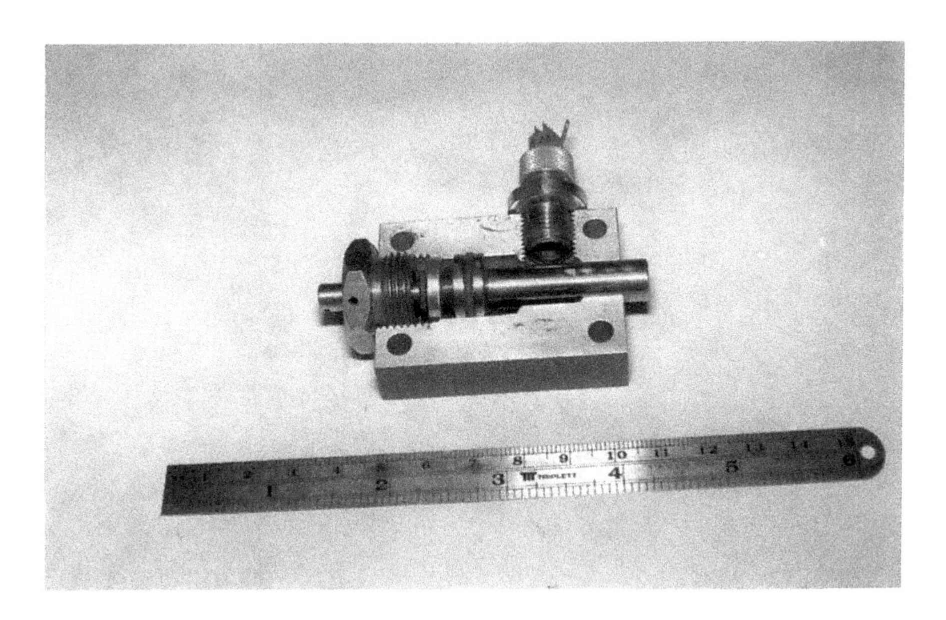

Figure 6.10. Piston motor-pin puller.

6.2.2 Piston-Driven Devices

These devices all use pyrotechnic or gas-generating materials to drive a piston, which, in turn, performs some kind of mechanical work. The simplest are the piston motors. Upon firing, the piston is driven forward. The major hazard from this device would be damage caused by the piston. Some piston motors can throw the piston clear of the assembly, thereby creating a low-velocity missile hazard. Most piston motors have captive pistons; that is, the piston is checked and held in place at the end of its designed travel. Figure 6.9 shows some piston motors.

The pin puller is virtually the same as the piston motor except that upon actuation the piston is driven into the assembly, pulling the exposed end in. This is shown in Figure 6.10, a cutaway of a combination piston motor-pin puller.

The piston motor can be designed integrally into another assembly where the mechanical function is completed. The simplest of these devices is the guillotine cutter. This is used to cut cables, wires, harnesses, and parachute reefing lines. They vary in size depending upon their design use. Figure 6.11 shows various guillotine cutter assemblies.

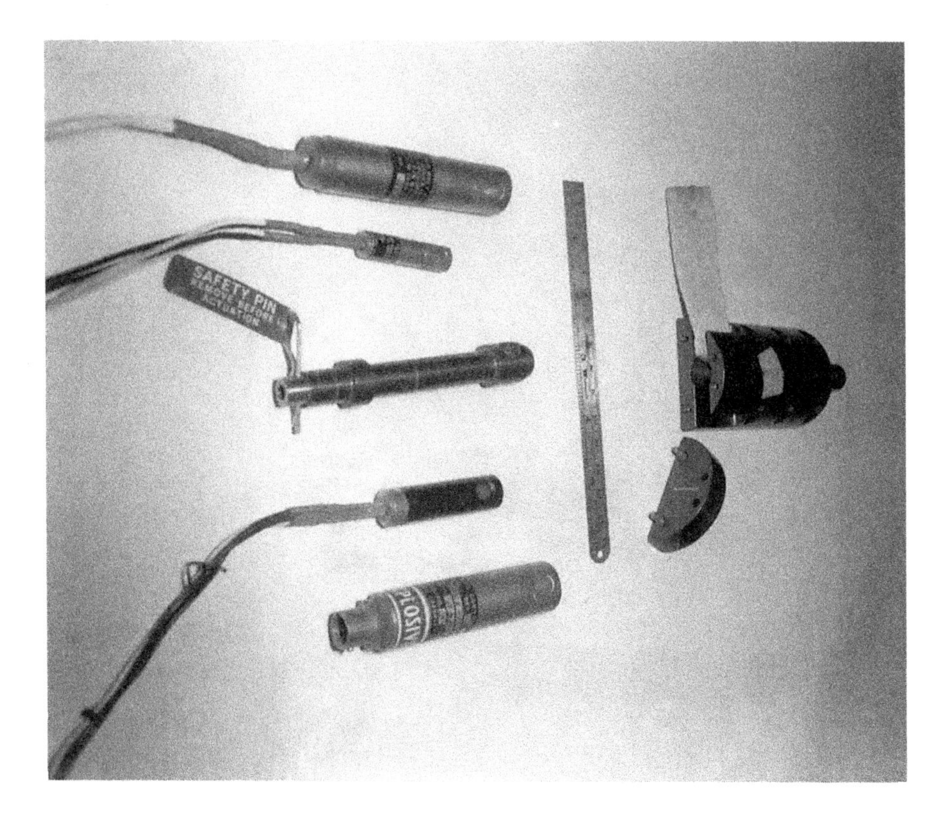

Figure 6.11. Guillotine cutter assemblies.

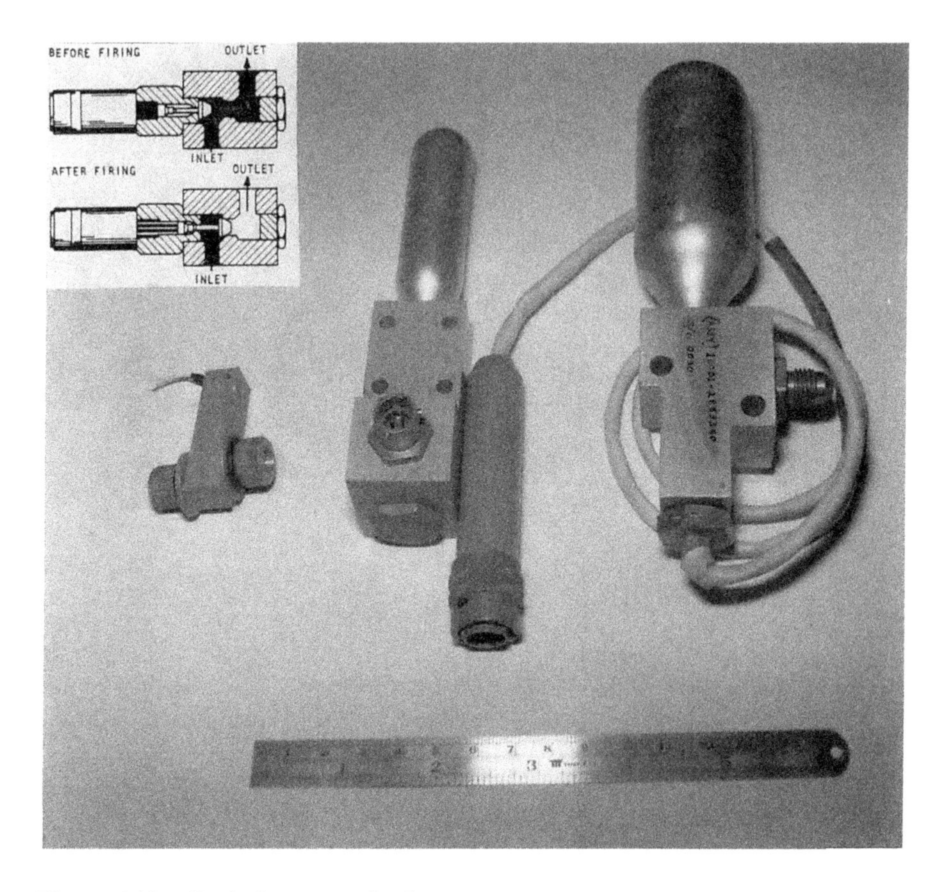

Figure 6.12. Explosive actuated valves.

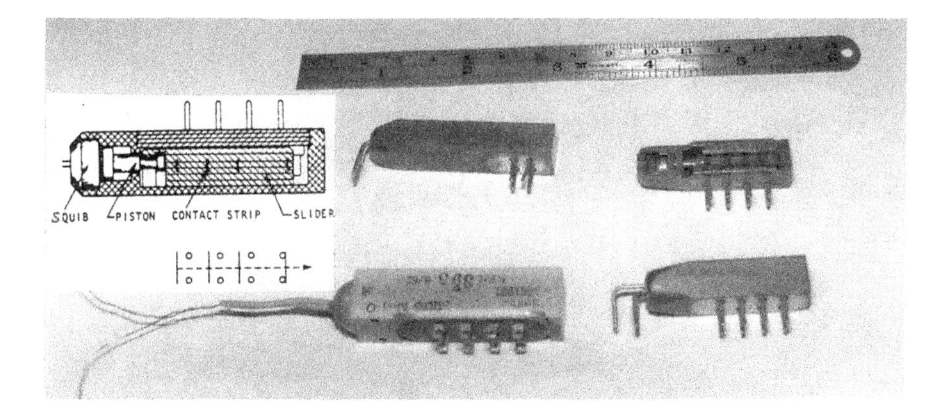

Figure 6.13. Explosive-driven switches.

Figure 6.14. Exploding nuts.

The piston is often used as the driver stem in a valve. Such valves are used in a wide variety of applications, ranging from underwater flotation recovery packages to emergency fuel dumps in space vehicles. Figure 6.12 shows several explosive-actuated valves.

The explosive electrical switches are totally self-contained devices like the explosive valves. Here, the piston drives the contact block in an electrical switch. This action can either "make" or "break" a circuit. Some switches are multi-

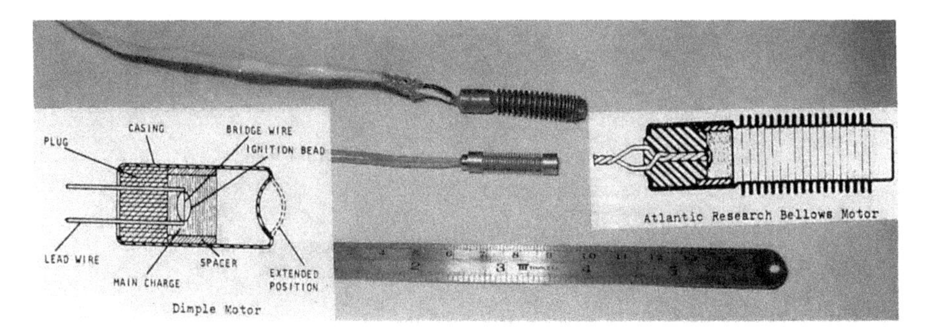

Figure 6.15. Dimple and bellows piston motors.

pole make and break; some others are simple single pole. Figure 6.13 shows a variety of explosive-driven switches.

Another explosive piston-driven device, used mainly on missile stage separation systems, is the exploding nut. Unlike the exploding bolt, the nut is not a detonating device. An explosively driven piston pushes a serrated nut through small tapered ramps. This causes the nut to expand and thereby releases the bolt, to which it is attached. Figure 6.14 shows four sizes of explosive nuts. The two center explosive nuts are shown after having been fired.

Other versions of the piston motor are the dimple motor and bellows motor. These are shown in Figure 6.15.

6.3 Related Reading

1. *Blaster's Handbook*, E. I. Dupont de Nemours and Co. (1969).
2. Brauer, K. O., *Handbook of Pyrotechnics*, Chem Pub. (1974).

CHAPTER 7

Classification, Transportation, and Storage of Explosives

In this chapter we shall look at some of the many regulatory and safety issues that affect and control the way we work with explosives. We shall start the chapter with the subject of classification of explosives and explosive devices into various categories that depend upon their initiation sensitivity. This is done up front because much of what follows depends upon which category of explosives we are working with. We shall then follow with regulatory requirements for the transportation and storage of explosives. In the next chapter are discussed explosives facilities, operations, and firing, and then we finish with a brief look at license and permit requirements in the United States and penalties imposed for violations of the laws and statutes dealing with explosives.

7.1 Explosives Classification

The basis for the classification of explosives throughout the world is based upon conventions developed by the United Nations Committee of Experts on the Transport of Dangerous Goods. This United Nations Committee (UNO) has developed a two-volume set of books (often referred to as the "orange books") pertaining to the transportation of dangerous materials.

Volume 1, *Recommendations on the Transport of Dangerous Goods*, cover principles of classification and definitions of classes, listings of the principal dangerous materials, general packing requirements, testing procedures, marking, labeling, placarding, and transport documents. There are also special recommendations related to particular classes of materials that do not apply to dangerous goods in bulk, but that are subject to special requirements in most countries.

Volume 2, *Tests and Criteria*, gives specific procedures and defines the tests required to determine the classification of an explosive or explosive device and the packaging proper for its transportation.

These volumes, along with several other pertinent references on this subject, are listed in the bibliography at the end of this chapter.

The transportation of dangerous materials is regulated in order to prevent accidents to personnel and property. However, the regulations are designed so as not to impede the movement of dangerous materials other than those that are too dangerous to be accepted for transporting. The regulations are addressed to all modes of transport. The classification of dangerous materials by type of risk involved has been drawn up to meet technical conditions while at the same time minimizing interference with existing regulations. It should be noted that the numerical order of the classes does not necessarily indicate the degree of danger. The recommended definitions of hazard classes in the UNO system are to indicate which materials are dangerous and in which class, according to their specific characteristics. Used with the list of dangerous materials found in Volume 1, the definitions provide guidance to those who have to use such regulations.

7.1.1 Hazard Class

The United Nations system (UNO) divides hazardous materials into nine (9) classes for the purpose of determining the degree of risk in shipping and transport. The order and number of the class are not meant to imply the degree of risk or danger. These classes are:

Class 1—Explosives
Class 2—Gases
Class 3—Flammable liquids
Class 4—Flammable solids
Class 5—Oxidizing substances, organic peroxides
Class 6—Toxic and infectious substances
Class 7—Radioactive materials
Class 8—Corrosive substances
Class 9—Miscellaneous dangerous substances and articles

Note that the information in this introductory text should not be used for determining the classification of a particular explosive or explosive article for the purpose of transporting of hazardous materials. This text is designed to introduce the reader to the "Recommendation on the Transport of Dangerous Goods" and the UNO system.

The nine classes of hazardous materials do not all pertain to explosives; they are covered in this text briefly to show the complexity of transporting these materials. For the purpose of this text, we will only consider Class 1, Explosives.

7.1.2 Hazard Division

Explosive substances and articles in Class 1 are assigned to one of six divisions, depending on the type of hazard they present. An explosive substance is defined as a solid or liquid substance (or mixtures of substances) that is in itself capable by chemical reaction of producing gas at such a temperature and pressure and at such a speed as to cause damage to the surroundings. Pyrotechnic substances are included even when they do not evolve gases. A pyrotechnic substance is defined as a substance, or mixtures of substances, designed to produce an effect by heat, light, sound, gas, smoke, or a combination of these as a result of a nondetonating self-sustaining exothermic reaction. An explosive article is defined as an article containing one or more explosive substances.

Division 1.1. *Substances and articles that have mass explosive hazard.* (A mass explosive is one that affects almost the entire load virtually instantaneously.)

Division 1.2 *Substances and articles that have a projection hazard but not a mass explosion hazard.*

Division 1.3 *Substances and articles that have a fire hazard and either a minor blast hazard or a minor projection hazard or both, but not a mass explosion hazard.*

This division comprises substances and articles that give rise to considerable radiant heat, or that burn one after another, producing minor blast or projection effects or both.

Division 1.4 *Substances and articles that present no significant hazards.*

This division comprises substances and articles that present only a small hazard in the event of ignition during transport. The effects are largely confined to the package, and no projection of fragments of appreciable size or range is to be expected. An external fire should not cause instantaneous explosion of the entire contents of the package.

Division 1.5 *Very insensitive substances and articles that have a mass explosion hazard.*

This division comprises substances that have a mass explosion hazard but are so insensitive that there is very little probability of ignition or transition from burning to detonation under normal conditions of transport. However, the probability of a transition from burning to detonation is greater when large quantities are carried in a ship.

Division 1.6 *Extremely insensitive articles that do not have a mass explosion hazard.*

This division comprises articles that contain only extremely insensitive detonating substances and demonstrate a negligible probability of accidental initiation or propagation. The risk from articles in Division 1.6 is limited to the explosion of a single article.

Class 1 explosives are unique in that the type of packaging frequently has a decisive effect on the hazard and therefore on the assignment to a particular division. The correct division is determined by use of the methods and tests described in Volume 2 of the orange books.

7.1.3 Compatibility Grouping

In addition to the Class and Division of an explosive, there is a third classification, and that is into compatibility groups. Explosive substances and articles are grouped according to whether they can be safely transported in the same unit load or stored together in the same magazine or bunker.

There are thirteen (13) compatibility groups, each designated by a letter of the alphabet. They are defined as follows:

A. Primary explosive substance.
B. Article containing a primary explosive substance and not containing two or more effective protective features. Some articles, such as detonators for blasting and detonator assemblies for blasting and primers, cap-type, are included, even though they do not contain primary explosives.
C. Propellant explosive substance or other deflagrating explosive substance or article containing such explosive substance.
D. Secondary detonating explosive substance or black powder or article containing a secondary detonating explosive substance, in each case without means of initiation and without a propelling charge, or article containing a primary explosive substance and containing two or more effective protective features.
E. Article containing a secondary detonating explosive substance, without means of initiation, with a propelling charge (other than one containing a flammable liquid or gel or hypergolic liquids).
F. Article containing a secondary detonating explosive substance with its own means of initiation, with a propelling charge (other than one containing a flammable liquid or gel or hypergolic liquids) or without a propelling charge.
G. Pyrotechnic substance, or article containing a pyrotechnic substance, or article containing both an explosive substance and an illuminating, incendiary, tear- or smoke-producing substance (other than a water-activated article or one containing white phosphorous, phosphides, a pyrophoric substance, a flammable liquid or gel, or hypergolic liquids).
H. Article containing both an explosive and white phophorous.
J. Article containing both an explosive substance and a flammable liquid or gel.
K. Article containing both an explosive substance and a toxic chemical agent.
L. Explosive substance or article containing an explosive substance and presenting a special risk (e.g., due to water activation or presence of hypergolic liquids, phosphides, or a pyrophoric substance) and needing isolation of each type.
N. Articles containing only extremely insensitive detonating substances.
S. Substance or article so packed or designed that any hazardous effects arising from accidental functioning are confined within the package unless the package has been degraded by fire, in which case all blast or projection effects are limited to the extent that they do not significantly hinder or prohibit fire

fighting or other emergency response efforts in the immediate vicinity of the package.

These definitions are those recommended by the UNO Panel in the orange book. The U.S. Departments of Defense and Energy have slightly different definitions of some of the groups because of their special needs in regard to weapons and munitions, but the spirit and intent of the orange book definitions are maintained.

Combining the class, division, and group defines the classification of an explosive substance or article. The classification for a particular item is stated as "class.division-group," for example: 1.1-D or 1.3-F.

Several of the references listed in the bibliography at the end of this chapter give lists of numerous examples of the various substances and articles in each of the divisions and groups.

It would be ideal if only one type of material or article were shipped or stored in any one vehicle or storage structure, but that is impractical. Therefore, some compatibility groups may be mixed within the same unit shipping load or in the same storage structure, but not within the same unitized pallet. Commercial practice in the United States, as dictated by the UNO recommendations and the U.S. Code of Federal Regulations (CFR), is slightly different in regard to groups mixing in storage from either the U.S. Department of Defense or Energy. The differences in mixed storage regulations among these three entities are due to the special needs and differences in the particular materials and articles involved and the manner in which they are packaged. Tables 7.1, 7.2, and 7.3 show which groups may be mixed in storage for each entity, respectively. Before assigning any explosive material or article to any class/division/group or place it in mixed groupings, consult the appropriate source documents.

In these three tables some of the allowable group mixing is designated by an "x," others by a "z." The "z" means that these combinations are subject to

Table 7.1 UNO Compatability Group Mixing Chart

Groups	A	B	C	D	E	F	G	H	J	K	L	N	S
A	x												
B		x											x
C			x	z	z							z	x
D			z	x	z							z	x
E			z	z	x							z	x
F						x							x
G							x						x
H								x					x
J									x				x
K										x			x
L											z		
N			z	z	z							x	x
S		x	x	x	x	x	x	x	x	x		x	x

Table 7.2 U.S. DoD Compatability Group Mixing Chart

Groups	A	B	C	D	E	F	G	H	J	K	L	S
A	x	z										z
B	z	x										x
C			x	z	z		z					x
D			z	x	x							x
E			z	x	x							x
F						x						x
G			z				x					x
H								x				x
J									x			x
K										z		
L												
S	z	x	x	x	x	x	x	x	x			x

Table 7.3 U.S. DOE Compatability Group Mixing Chart

Groups	A	B	C	D	E	F	G	L	S
A	x	z							
B	z	x	z	z	z	z	z		x
C		z	x	x	x	z	z		x
D		z	x	x	x	z	z		x
E		z	x	x	x	z	z		x
F		z	z	z	z	x	z		x
G		z	z	z	z	z	x		x
L								x	
S		x	x	x	x	x	x		x

special caveats that are spelled out in great detail in the respective source documents.

7.2 Transportation of Explosives

The UNO recommendations in the orange book specify in detail the restrictions on containers, road vehicles, and rail wagons that may be used for explosives transport. These are quoted in the following: "Freight containers, vehicles and wagons should not be offered for the transport of explosive substances and articles of Class I, unless the freight container, vehicle or wagon is structurally serviceable as witnessed by current International Convention for Safe Containers (CSC) approval plate (applicable to freight containers only) and a detailed visual examination as follows:

1. Prior to loading a freight container, vehicle or wagon with explosives, it should be checked to ensure it is free of any residue of previous cargo and to ensure it is structurally serviceable and the interior floor and walls are free from protrusions.
2. ''Structurally serviceable'' means that the freight container, vehicle or wagon is free from major defects in its structural components, top and bottom side rails, top and bottom end rails, door sills and headers, floor crass members, corner post, and corner fittings in a freight container. Major defects are dents or bends in structural members greater than 19 mm in depth, regardless of the length; cracks or breaks in structural members; more than one splice in top or bottom end rails, or door headers or more than two splices in any one top or bottom side rail or any splice in a door sill or corner post; door hinges and hardware that are seized, twisted, broken, missing or otherwise inoperative, gaskets and seals that do not seal; freight containers, any distortion of the over-all configuration great enough to prevent proper alignment of handling equipment, mounting and securing on chassis, vehicles or wagons, or insertion into ship's cells.
3. In addition, deterioration in any component of the container, vehicle or wagon, regardless of the material of construction, such as rusted-out metal in side-walls or disintegrated fiberglass is unacceptable. Normal wear, however, including oxidization, slight dents and scratches and other damage that do not affect serviceability or the weather-tight integrity of the units is acceptable. For free-flowing powdery substances of 1.1C, 1.1D, 1.1G, 1.3C, and 1.3G and fireworks of 1.1G, 1.2G, and 1.3G, the floor of the freight container should have a non-metallic surface or covering.''

7.2.1 Vehicle Inspection

All vehicles that are used for the movement of explosives must meet special equipment requirements. A source of these requirements may be found in the DOE Explosives Safety Manual, DOE/EV06194. In addition to these specifications, each vehicle (including forklifts and other powered equipment) must be inspected daily before being used. A record of the inspection should be maintained. Figure 7.1 shows a typical daily vehicle inspection checksheet.

7.2.2 Load Tie Down

All explosives carried on any vehicle must be properly secured. Securing the load to the transport vehicle is accomplished by a variety of methods, including blocking and bracing with wood, tying down the boxes or containers with nylon load binders, chains, and in some cases, banding groups of containers together. Small items may be packed into approved explosive transport cases, which are secured to and part of the transport vehicle.

DAILY VEHICLE INSPECTION
— CHECK SHEET —

(Inspect each item listed at the start of each scheduled work day before using the vehicle. Check each item after completion and note any defects in the remarks column. Safety deficiencies must be corrected before use. Return this inspection upon completion to the unit supervisor.)

VEHICLE TYPE **NUMBER**
Date ____________________ ____________________

No.	EQUIPMENT CHECKS	REMARKS
1	GAS CAP	
2	LEAKS	
3	TIRES	
4	TAILGATES/DOORS/LOCKS	
5	SIDEBOARDS	
6	MIRRORS-LEFT/RIGHT	
7	FIRE EXTINGUISHERS (2)	
8	EXPLOSIVE PLACARDS (4)	
9	WHEEL CHOKES	
10	CARGO STRAPS	
11	CARGO TIE-DOWN RINGS	
12	CARGO BED CLEAN	
13	WINDSHIELD WIPERS	
14	ALL LIGHTS	
15	FUEL (1/2+), OIL, WATER	
16	REAR DOOR ALARM	
17	RADIO	
18	CAB INTERIOR CLEAN	
19	SEAT BELTS	
20	ELECTRICAL WIRING	
21	EXHAUST SYSTEM	
22	BRAKES	

INSPECTED BY: ________________, **SUPERVISOR:** ________________

Figure 7.1. Typical explosive vehicle daily inspection check sheet.

Table 7.4 Placards Required for Vehicles Transporting Dangerous Goods

Category of Material	Placard Name	Placard Design Spec (Per 49CFR-172)
1.1	EXPLOSIVES 1.1	172.522
1.2	EXPLOSIVES 1.2	172.522
1.3	EXPLOSIVES 1.3	172.522
1.4	EXPLOSIVES 1.4	172.523
1.5	EXPLOSIVES 1.5	172.524
1.6	EXPLOSIVES 1.6	172.525
2.1	FLAMMABLE GAS	172.532
2.2	NON-FLAMMABLE GAS	172.528
2.3	POISON GAS	172.540
3	FLAMMABLE	172.542
Combustible liquid	COMBUSTIBLE	172.544
4.1	FLAMMABLE SOLID	172.546
4.2	SPONTANEOUSLY COMBUSTIBLE	172.547
4.3	DANGEROUS WHEN WET	172.548
5.1	OXIDIZER	172.550
5.2	ORGANIC PEROXIDE	172.552

7.2.3 Placarding

All vehicles carrying explosives and or explosive articles must be placarded with the type of placards specified in Table 7.4.

7.3 Storage of Explosives

Storage facilities for explosives must be constructed to meet a number of specific requirements, which are spelled out in detail in the various regulatory source documents listed in the bibliography at the end of this chapter. In addition, all storage facilities should have the fire symbol posted in a location that is visible to an approaching vehicle. All walkin-type storage facilities should have placards on or near the entry doors specifying personnel limits and the compatability groups. Vegetation and other combustibles around storage facilities should be controlled to minimize any possible grass, brush, forest, or trash fires. All storage facilities should have fire extinguishers with a rating of 2A: 20BC (2A wood type; B, flammable liquid; C, electrically nonconductive), and all personnel handling explosives must know the location of the fire extinguishers in the storage facilities. All explosives should be properly packaged and labeled while in storage, with the exception of day-use magazines. Personnel are not allowed to work in storage magazines or igloos other than to open and inspect the contents of an individual package. A Safe Operating Procedure should be required for each facility to cover all aspects of facility use, including the items mentioned and any specific requirements.

7.3.1 Types of Storage Magazines

A magazine is any building or structure, except an operating building, used for the storage of ammunition and explosive. The definitions of types, construction, and storage limitations for magazines are somewhat different in commerce and the military because of the nature of the materials and articles stored and the logistics of the storing entity.

7.3.1.1 U.S. Commercial Storage

For the purpose of storage, the U.S. regulations (27 CFR Part 55) break explosives into three types, each of which must be stored in separate magazines. These types are (as quoted from 27 CFR Part 55):

(a) High Explosives. Explosive materials which can be caused to detonate by means of a blasting cap when unconfined (for example, dynamite, flash powders, and bulk salutes).
(b) Low Explosives. Explosive materials which can be caused to deflagrate when confined, (for example, black powder, safety fuses, igniters, igniter cords, fuse lighters, and "special fireworks" as defined by U.S. DOT regulations in 49 CFR Part 173, except for bulk salutes).
(c) Blasting Agents (for example, ammonium nitrate-fuel oil and certain water gels)."

Five types of magazines are specified and defined in 27 CFR Part 55:

(a) Type 1, permanent magazines for the storage of high explosives.
(b) Type 2, mobile and portable indoor magazines for the storage of high explosives.
(c) Type 3, portable outdoor magazines for the temporary storage of high explosives while attended (for example, a "day box").
(d) Type 4, magazines for the storage of low explosives (under certain limitations blasting agents as well as detonators that will not mass detonate may be stored in Type 4 magazines).
(e) Type 5, magazines for the storage of blasting agents.

The construction requirements for all of the above magazine types are given in detail in the same regulations. The requirements include walls, doors, floors, rooves, and foundations. In essence, these design requirements provide for fire safety, bullet proofing, and security. In regard to the latter, locks, hasps, and hinges are also specified.

7.3.1.2 Military Magazines

There are essentially two basic types of military magazines, earth covered and above ground.

1. Earth-covered types are igloos, steel arch, stradley, hillside, and subsurface. These are considered "preferred magazines." In addition to security and safety from fires and explosions, the earth covering provides thermal insulation and temperature stability.
2. Above-ground types are constructed with concrete floors and walls made of concrete, tile, or block. The tile or block in some magazines may be filled with sand depending on the intended storage.

General specifications provide that:

1. No wooden floors may be used,
2. Construction must allow for ease of cleaning and no accumulation of explosive dusts, powders, etc.,
3. Doors are fire resistant, insulated, and tight fitting,
4. Openings for ventilation must be screened to prevent entry of insects, rodents, and reptiles,
5. No heating is provided (except for special requirements, demanded by the particular munition being stored).

7.3.2 Housekeeping and Fire Protection

Vegetation around storage magazines should be maintained to minimize the potential damage to the facilities and for the safety of personnel. A firebreak of 15 meters in width and free from all combustible materials should be maintained around all explosive facilities.

Regulations require that "magazines should be kept clean, dry, and free of grit, paper, empty packages, and containers, and rubbish. Brooms and other utensils used in the cleaning and maintenance of magazines must have no spark producing metal parts, and may be kept in magazines. Floors stained by leakage from explosive materials are to be cleaned according to instructions of the explosive manufacturer. When any explosive material has deteriorated it is to be destroyed in accordance with the advice or instructions of the manufacturer. The area surrounding magazines is to be kept clear of rubbish, brush, dry grass, or trees (except live trees more than 10 feet tall), for not less than 25 feet in all directions. Volatile materials are to be kept a distance of not less than 50 feet from outdoor magazines. Living foliage which is used to stabilize the earthen covering of a magazine need not be removed."

7.3.3 Lighting

In general, storage magazines are not lighted. For temporary lighting, battery-activated safety lights or lanterns may be used. For permanent lighting installations the electric lighting used in any explosive storage magazine must meet the standards prescribed by government regulations for the conditions present in the magazine at any time. In the United States those standards are provided by the "National Electrical Code" (National Fire Protection Association, NFPA 70-81). All electrical switches should be located outside the magazine and also meet the standards of the prescribed code.

7.3.4 Storage and Storage Containers

Explosive items should be properly packaged when stored. The containers should be constructed so that they will not leak and will protect their contents from excessive movement, contamination, or spillage during handling. The containers should be constructed of, or lined with, nonabsorbent materials that are compatible with the explosive contents. Glass containers should not be used except for small samples in cases where the explosive will react with other materials. When storing explosive powders, or explosive dust products, do not use containers with seams or rivets. Plastic conductive containers should be used.

Storage containers should not be placed against the walls of the magazine. They should be placed so that ventilation is not inhibited. It is permissible to stack similar items in storage. When stacking such items, provisions should be made so that air is free to circulate to all parts of the stack. Limits on the stack height will depend on the container and the stability of the stacks. If shelves are provided in magazine, explosives items should not be placed higher than 2 meters from the floor. Aisles in magazines should be sufficiently wide to accommodate inspection, inventory, sampling, and material handling operations on the stored explosives containers. Explosive storage containers should be placed so that marking and labeling are clearly visible and that the materials can be easily inspected and counted. Containers should be sealed, or at least closed, in storage.

All explosive items that are stored should be labeled. The storage container should, at a minimum, show the following information:

1. A storage tag or inventory control number,
2. Item name,
3. Brief description of item,
4. Weight of explosive, and
5. Class/division/compatibility group.

If there are several separate containers with the same contents, each container should have all the above information. However, if all the containers are on one pallet and are bound together, one information packet is sufficient.

7.3.5 Quantity-Distance Criteria

In order to ensure the safety of personnel, facilities, and the general public, explosive storage magazines must be kept at some minimum distance from occu-

Table 7.5 Commercial Q-D Criteria for Storage of High Explosives, Barricaded

	Distances (in feet)			
Quantity of Explosives (pounds)	Inhabited Buildings	Public Highways (fewer than 3000 vehicles per day)	Passenger Railways & Public Highways (more than 3000 vehicles per day)	Separation of Magazines
5	70	30	51	6
10	90	35	64	8
20	110	45	81	10
30	125	50	93	11
40	140	55	103	12
50	150	60	110	14
75	170	70	127	15
100	190	75	139	16
200	235	95	175	21
300	270	110	201	24
400	295	120	221	27
500	320	130	238	29
600	340	135	253	31
700	355	145	266	32
800	375	150	278	33
900	390	155	289	35
1000	400	160	300	36
2000	505	185	378	45
3000	580	195	432	52
4000	635	210	474	58
5000	685	225	513	61
6000	730	235	546	65
7000	770	245	573	68
8000	800	250	600	72
9000	835	255	624	75
10000	865	260	645	78
20000	975	290	813	98
30000	1130	340	933	112
40000	1275	380	1026	124
50000	1400	420	1104	135
60000	1515	455	1173	145
70000	1610	485	1236	155
80000	1695	510	1293	165
90000	1760	530	1344	175
100000	1815	545	1392	185
200000	2030	610	1755	285
300000	2275	690	2000	385

Table 7.6 Commercial Q-D Criteria for Storage of High Explosives, Unbarricaded

Quantity of Explosives (pounds)	Distances (in feet)			
	Inhabited Buildings	Public Highways (fewer than 3000 vehicles per day)	Passenger Railways & Public Highways (more than 3000 vehicles per day)	Separation of Magazines
5	140	60	102	12
10	180	70	128	16
20	220	90	162	20
30	250	100	186	22
40	280	110	206	24
50	300	120	220	28
75	340	140	254	30
100	380	150	278	32
200	470	190	350	42
300	540	220	402	48
400	590	240	442	54
500	640	260	476	58
600	680	270	506	62
700	710	290	532	64
800	750	300	556	66
900	780	310	578	70
1000	800	320	600	72
2000	1010	370	756	90
3000	1160	390	864	104
4000	1270	420	948	116
5000	1370	450	1026	122
6000	1460	470	1092	130
7000	1540	490	1146	136
8000	1600	500	1200	144
9000	1670	510	1248	150
10000	1730	520	1290	156
20000	1950	580	1626	196
30000	2000	680	1866	224
40000	2000	760	2000	248
50000	2000	840	2000	270
60000	2000	910	2000	290
70000	2000	970	2000	310
80000	2000	1020	2000	330
90000	2000	1060	2000	350
100000	2000	1090	2000	370
200000	2030	1220	2000	570
300000	2275	1380	2000	770

pied buildings, public highways, and passenger railways. Both civilian and military regulations have been established that specify these distances based upon the type and quantity of explosives being stored. These regulations are in the form of quantity-distance (Q-D) tables. The tables are based upon safe distances from a given size explosion to prevent injury or damage due to blast pressure, high-velocity fragments, or debris.

7.3.5.1 Commercial Tables

For the storage of commercial explosives, different tables are given in 27 CFR Part 55 for the different types of explosives (high explosives, low explosives, and blasting agents). These tables consider safe distances for blast and debris but not for high-velocity metal fragments, since it is assumed that commercial explosives are not stored in metal casings (such as bombs and other military ordnance). Tables 7.5 and 7.6 are used for magazines storing high explosives with and without barricades; Table 7.7 is used for low explosives.

In addition to these tables, Q-D tables are also given in 27 CFR Part 55 for storage and operations facilities involved with the manufacture of fireworks, but these are not included here.

7.3.5.2 Military Tables

The Q-D tables required for the storage of military explosives and ordnance are different from the commercial tables in that they account for safe distance

Table 7.7 Commercial Quantity-Distance Criteria for Storage of Low Explosives

Pounds of Low Explosives		Distances (feet)		
Over	Under	From Inhabited Building	From Public Railroad and Highway	From Above-Ground Magazine
0	1,000	75	75	50
1,000	5,000	115	115	75
5,000	10,000	150	150	100
10,000	20,000	190	190	125
20,000	30,000	215	215	145
30,000	40,000	235	235	155
40,000	50,000	250	250	165
50,000	60,000	260	260	175
60,000	70,000	270	270	185
70,000	80,000	280	280	190
80,000	90,000	295	295	195
90,000	100,000	300	300	200
100,000	200,000	375	375	250
200,000	300,000	450	450	300

Table 7.8 DoD Quantity-Distance Table for Storage of Class/Division 1.1 Explosives

NEW (net explosive weight) (pounds)	Distance to Inhabited Buildings (in feet) from: Earth-Covered Magazines, From Front or Side	Distance to Inhabited Buildings: Earth-Covered Magazines, From Rear	Distance to Inhabited Buildings: Other Above-Ground Storage	Distance to Public Traffic Route (in feet) from: Earth-Covered Magazines, From Front or Side	Distance to Public Traffic Route: Earth-Covered Magazines, From Rear	Distance to Public Traffic Route: Other Above-Ground Storage
1–200	500	250	1250	300	150	750
200–500	700	"	"	420	150	"
500–35000	1250	1250	"	750	750	"
35000	"	"	1310	"	"	785
40000	"	"	1370	"	"	820
50000	1290	"	1475	775	"	885
60000	1370	"	1565	820	"	940
70000	1440	"	1650	865	"	990
80000	1510	"	1725	905	"	1035
90000	1570	"	1795	940	"	1075
100000	1625	"	1855	975	"	1115
125000	1910	1480	2115	1165	890	1270
150000	2175	1805	2350	1305	1085	1410
175000	2435	2135	2565	1460	1280	1540
200000	2680	2470	2770	1610	1480	1660
250000	3150	3150	3150	1890	1890	1890
300000	3345	3345	3345	2005	2005	2005
350000	3525	3525	3525	2115	2115	2115
400000	3685	3685	3685	2210	2210	2210
450000	3830	3830	3830	2300	2300	2300
500000	3970	3970	3970	2380	2380	2380

for high-velocity metal fragments and therefore include larger initial distances. This is seen in Table 7.8, which is for magazines storing Class/Division 1.1 explosives.

Because the military often has facilities where various magazines of different types may be in the same area, there is a special table for intermagazine distances based on type of structure as well as type and quantity of explosive. This is shown in Table 7.9

In addition to these tables, the military also specifies Q-D tables for: military aircraft parking areas; airfields, heliports, and seadromes; pier and wharf facilities; and liquid propellants.

Table 7.9 DoD Guide for Intermagazine Quantity-Distance Criteria for Storage of Class/Division 1.1 Explosives (numbers in table are values of K in the distance/quantity scaling equation, $D = KW^{1/3}$, where D is the distance in feet and W is the weight of explosives in pounds)

		Standard Earth-Covered Magazine				Above-Ground Magazine	
From:	To:	Side	Rear	Front, Unbarricaded	Front, Barricaded	Unbarricaded	Barricaded
Standard earth-covered magazines	Side	1.25	1.25	2.75	2.75	6.0	4.5
	Rear	1.25	1.25	2.0	2.0	6.0	4.5
	Front, unbarricaded	2.75	2.0	11.0	6.0	11.0	6.0
	Front, barricaded	2.75	2.0	6.0	6.0	6.0	6.0
Above-ground magazines	Unbarricaded	4.0	4.0	11.0	6.0	11.0	6.0
	Barricaded	4.0	4.0	6.0	6.0	6.0	6.0

7.3.5.3 Table for Underground Utilities

Besides providing protection for personnel and above-ground facilities, care must be taken in locating explosive magazines that may be in the vicinity of underground electrical cabling, service installations, and piping. Table 7.10 gives the recommended Q-D requirements for these.

Table 7.10 Quantity-Distance Separation for the Protection of Underground Service Installations

Quantity of Explosives	Distance (feet)
100	80
200	"
500	"
1,000	"
2,000	"
5,000	"
10,000	"
20,000	85
50,000	110
100,000	140
250,000	190

7.4 Related Reading

1. *United Nations Recommendations on the Transport of Dangerous Goods*, ST/SG/AC.10/1.
2. *United Nations Recommendations on the Transport of Dangerous Goods—Tests and Criteria*, ST/SG/AC.10/11.
3. Code of Federal Regulations, 27 CFR Part 55.
4. Code of Federal Regulations, 49 CFR Parts 106–180.
5. *Dangerous Goods Regulations*, International Air Transport Association (IATA).
6. *Explosives Law and Regulations*, ATF P 5400.7, U.S. Bureau of Alcohol, Tobacco, and Firearms.
7. *U.S. Department of Defense Ammunition and Explosives Safety Standards*, DoD 6055.9-STD.
8. *U.S. Army Material Command Safety Manual*, AMC-R 385-100.
9. *U.S. Air Force Explosive Safety Manual*, AFR 127-100.
10. *U.S. Department of Defense Contractors Safety Manual for Ammunition and Explosives*, DoD 4145.26-M.
11. *U.S. Department of Energy Explosives Safety Manual*, DOE/EV/06194.
12. *Manufacture, Transportation, Storage, and Use of Explosives*, NFPA 495, National (U.S.) Fire Prevention Association.
13. *Safety Library Publications* from the Institute of Makers of Explosives (IME), 1120 19th Street NW, Suite 310, Washington, DC 20036.
    - 13.1 Construction Guide for Storage Magazines (8/93).
    - 13.2 American Table of Distances (6/91).
    - 13.3 Warnings and Instructions for Consumers in Transporting, Storing, Handling and Using Explosive Materials (3/92).
    - 13.4 Handbook for the Transportation and Distribution of Explosive Materials (6/93).
    - 13.5 Safety in Transportation, Storage, Handling and Use of Explosive Materials (3/87).
    - 13.6 Recommendations for the Safe Transporation of Detonators in the Same Vehicle with Certain Other Explosive Materials (5/93).
    - 13.7 Generic Guide for the Use of IME 22 Container (1/95).

CHAPTER

8

Explosive Facilities and Explosives Operations

This chapter gives us a brief look into the general requirements for explosives facilities and equipment, general explosives operations, firing explosives operations, and some of the licensing requirements imposed by the government.

Because of the potentially disastrous effects of an accident or explosion, very stringent requirements are placed on facilities in which explosives and explosive articles are manufactured and tested. In addition to international requirements (which were seen in the previous chapter), the laws and regulations of the nation are most often supplemented by local governments. It is very important, before embarking on any explosive operation, to consult those laws and regulations that will certainly include those for worker health and safety, public safety, air pollution, water pollution (both surface and underground), land use, and noise and ground vibration control.

8.1 Explosive Facilities

The design of explosive facilities is driven by where and of what type and what quantity of explosives will be present. In production, usually the explosives are inside the structure. In testing, often the explosives are outside the structure.

8.1.1 General Structural Considerations

An accident at an explosive facility could occur indoors, in which case the facility should be designed to protect people and property in adjacent rooms and adjacent buildings. Many facilities utilize design features such as blowout walls or roofs. In such buildings, the inner walls are very strong; in the event of an explosion these walls would protect people in the building but not in that room. If blowout walls are utilized, then access to the outside of those walls must be restricted while explosives are in process behind them. Each building or laboratory or room containing explosives or any material that may be hazardous to operating personnel should have at least two exits. If the building or room is occupied by not more than two people, one exit is usually acceptable. All exits should be at least 30 in. in width, and where possible, the exits should be at the opposite ends of the building or room. The explosive hazard should not be located between the operator and the exit. Whenever possible, the exit should be directly to the outside of the building and not into a hallway or another room.

In facilities where the explosive work is outside a particular building, then the outer walls and rooves of that building should be strong enough to resist blast insult and protect people and property inside. Conventional structures are designed to withstand roof snow loads of 30 pounds per square foot and wind loads of 100 miles per hour. These loads equate to only 0.2 pounds per square inch, considerably less than the loads to which a blast-resistant structure is designed. Design maximum air blast overpressure for an unbarricaded line-of-site uninhabited structure is typically 3.5 psi and for an inhabited one is 0.9 to 1.2 psi.

8.1.2 Barricades, Shields, and Test Chambers

Where the explosion hazard is of greater magnitude than can be economically designed into a particular building, then barricades are employed to reduce the effects on the structure.

8.1.2.1 Barricades

A barricade is an approved obstacle, natural or artificial, constructed to limit the effects of an explosion on nearby buildings or personnel. There are two types of barricade, natural and manmade. The natural barricades can be forests, hills, or valleys, and manmade can be any structure such as concrete, dirt, rocks, wood, etc. Air blast is just one concern when designing structures that must be protected from an explosion. Fragments from an explosion can be as much or more of a hazard than the air blast. Barricades must be constructed to protect against both primary and secondary fragments. Primary fragments are defined as small high-velocity particles from castings or containers that surround the explosive's source. Secondary fragments are heavy low-velocity particles that are formed from pieces of equipment and structures near the explosive source. Primary fragments usually have velocities on the order of thousands of feet per second, and secondary fragments usually have velocities on the order of hundreds of feet per second.

When designing shields or barricades, the type of explosive must be considered to determine the amount of energy release by the heat of detonation. Also, if the calculations pertain to an enclosed test chamber, the complete chemical reaction (heat of detonation and heat of combustion) must be considered because the pressures in both reactions are additive. Detonating an explosive charge on the surface of the ground will generate both an air-blast and ground-shock wave, both of which must be considered when designing any type of explosive protection. A large percentage of high explosives are underoxidized; in these cases, approximately one-third of the total chemical energy available is released in the detonation. The remaining chemical energy is released slowly in a secondary reaction as the products mix with air and burn. This secondary reaction has only a slight effect on the blast-wave properties because it is much slower. This secondary reaction is known as a secondary fireball.

When an explosive detonates, temperatures of approximately 3000°C are produced. These gases cool rapidly as they expand. Temperatures of 400°F are enough to ignite most organic materials. Grass, brush, and debris should be kept to a minimum around all explosive facilities. If barricaded open-storage modules are being used, propagation of explosions by fire is possible. Also, work areas should be barricaded and far enough away to allow expanding gases to cool and decrease in pressure to a level the human body can tolerate.

8.1.2.2 Shields

Throughout the commercial industries and government laboratories that use explosives, the words *barrier*, *barricade*, and *shield* represent interchangeable concepts. However, the word *shield* is usually associated with the device used to protect an individual while working on a bench, or a scale when handling explosives. Normally, they are designed with a heavy nonsparking base that holds a clear plastic window (shield) approximately 2 ft by 2 ft by 1 in. thick. Explosive loading presses have either a metal or plastic sliding door that the operator will close when pressing a pellet or loading a component. Table 8.1 lists various types of safety shields and explosive limits for each shield.

8.1.2.3 Test Chambers

There are several types and styles of explosive test chambers, ranging from complete containment to containment of fragmentation only. Normally, explosive test chambers are designed to vent the gases and contain the fragments. However, some are designed to contain all the products from an explosion. Some explosive experiments require that the test engineer be near the test chamber when the explosive is fired. In this situation the test chamber has to be proof tested to 125% of the maximum anticipated pressure.

8.1.3 Electrical Requirements

The minimum requirements for electrical wiring in explosive facilities should meet the specifications of the *National Electrical Code* (NEC). All permanent

Table 8.1 Some Safety Shields for Explosives Laboratory Operations

Shield Type	Explosive Limit (grams)	Minimum Distance from Explosive (cm)
Leather gloves, jackets or coats, plastic face shields	0.05	0
3-mm tempered glass	0.05	8
7-mm polymethylmethacrylate clear sheet plastic (or equivalent)	2.5	15
20-mm polymethylmethacrylate clear sheet plastic (or equivalent)	10	15
15-mm laminated bullet-resistant glass	20	20
25.4-mm polycarbonate clear sheet plastic (or equivalent)	50	30

electrical equipment and wiring for areas containing explosives should conform to the standards of NEC for ''hazardous location'' Class I or Class II.

There are two basic types of wiring systems used in explosive facilities: *explosion-resistant* and *explosion-proof.*

1. The *explosion-resistant* systems (Class II locations) consists of conduit, receptacles, and switches that have seals in all connections so that explosive dust will not filter into the system. This type of system should be the minimum for most solid explosive facilities, especially those areas containing explosives dust, or explosives that may produce dust.
2. The *explosion-proof* systems (Class I locations) are those that will not propagate an internal explosion caused by infiltration of gaseous materials. They are not immune to damage caused by external explosions. These systems are designed using O-ring seals to minimize vapor penetration into the system. The conduit, receptacles, and switches have an internal volume such that if flammable vapors were to filter into the system and an electrical arc were generated, there would not be enough oxygen to react and cause an explosion. If the vapors were from an explosion and the vapors contained an explosive fuel and oxidizer mixture, then exposure to an electrical arc would not release sufficient explosive energy to destroy the fixture. The internal volume of this system is so small that the conduit, receptacles, and switches will contain the explosion. This type of system is used in hazardous locations where flammable vapors may be present. Flammable and explosive vapors can be produced by sublimation of some solid explosives and vaporization of most liquid explosives.

Beside the wiring systems of the buildings themselves, consideration must be given to the requirements of all electrical equipment in the facility. All equipment used in conjunction with bare explosive materials have special features that are different from analogous equipment used in nonexplosive areas. For example, all electrical wiring must meet the requirements of either explosion-proof or explosion-resistance standards. All recessed-head screws must have the socket filled with potting material to prevent the accumulation of explosive dusts; machines must use only vacuum chucks; and holding fixtures must be made of nonsparking materials. Equipment must be cleaned after each operation so that it will not become contaminated with explosives.

Ground fault circuit interrupters (GFCI) should be used in all areas where electrical shock may be a hazard, such as in wet and damp process areas and with machinery. The GFCI is a device to protect personnel from an electrical shock. GFCI protection should be provided in static grounded areas where personnel may come into contact with AC-powered electrical equipment. According to specifications, a GFCI will interrupt with a short-circuit current of 5 to 6 milliamperes in 2 to 5 seconds and 0.25 amperes in 30 to 40 milliseconds. However, examination of the acceptable operational envelope shown in Figure 8.1 reveals that GFCIs do not provide absolute protection against electrical shock and possibly no protection from high currents (in excess of a quarter ampere).

8.1.4 Control of Static Electricity

Positive steps must be taken to control or eliminate static electricity in work areas where explosive materials or explosive articles may be ignited from an electrostatic discharge (ESD). This includes ESD-sensitive propellants, pyro-

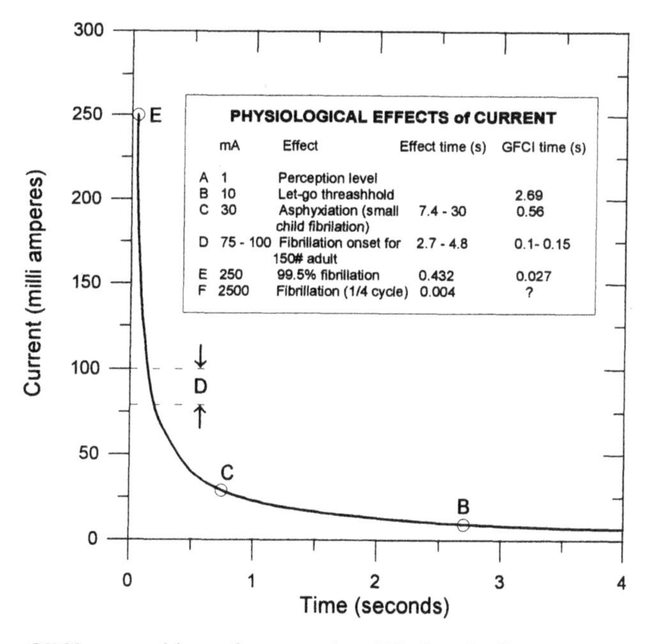

PHYSIOLOGICAL EFFECTS of CURRENT

	mA	Effect	Effect time (s)	GFCI time (s)
A	1	Perception level		
B	10	Let-go threashhold		2.69
C	30	Asphyxiation (small child fibrilation)	7.4 - 30	0.56
D	75 - 100	Fibrillation onset for 150# adult	2.7 - 4.8	0.1- 0.15
E	250	99.5% fibrillation	0.432	0.027
F	2500	Fibrillation (1/4 cycle)	0.004	?

Figure 8.1. GFCI acceptable performance (per UL Standard).

technics, solvents, vapors, and flammable gases. Work areas where ESD-sensitive explosives are handled should be equipped with conductive floors, conductive chairs, and conductive workbench tops, and all should be connected to a common ground. The resistance of a floor should be more than 5,000 ohms in areas with 110-volt service and 10,000 ohms with 220-volt service, and less than 1,000,000 ohms in all areas from floor to the ground plane. Test meters used for testing conductive surfaces generally operate with an open-circuit voltage of 500 volts DC and a short-circuit current of 2.5 milliamperes with an internal resistance of approximately 200,000 ohms. Each electrode should weigh about 2.5 kg and be 6 to 7 cm in diameter. The conductive surface of the electrodes should be aluminum or aluminum foil, 1 to 2.5 mm thick, backed with a layer of 40 to 60 durometer rubber.

All explosive structures should have their own individual ground and should be isolated from any of the building's electric system grounds. If the need arises to work with grounded explosives in an area where explosives are not normally handled, do not connect the ground to gas, steam, air, dry sprinkler systems, or air terminals of lighting protection systems. Static grounds may be connected to water pipes, ground cones, buried copper plates, or driven ground rods. The normal resistance between all conductive parts of equipment and ground should not exceed 25 ohms. All grounding systems should be visually inspected and tested for resistance to ground periodically, depending on the system and management requirements.

8.1.5 Lightning Protection

All governments require that lightning protection systems should be provided on all facilities used for storage, processing, handling, testing, or other work with explosives. Specifications for such systems can be found in NFPA 780, "Lightning Protection Code," Appendix L ("Protection of Structures Housing Explosives"). Examples of a number of different systems and system configurations can be found in DoD 6055.9-STD (listed in the previous chapter).

Lightning protection systems should be visually inspected twice a year for evidence of corrosion, broken wires and/or connectors, or any other problem that would impair the effectiveness of the system. Testing of systems should be conducted annually. Testing should be done only with instruments specially designed for that purpose and designed for earth-grounding systems.

8.2 Explosives Operations

Explosives operations involve many different activities, each with its own particular hazards and specialized equipment. A list of some of the operations commonly performed includes:

Synthesis
Washing

Screening
Wet and dry milling
Blending, formulating, and mixing
Heating and drying
Punch and die pressing
Isostatic pressing
Extruding
Machining
Melting and casting
Assembly and disassembly
Packaging
In-plant movement

Since each of these operations will involve specialized equipment, the maintenance and cleaning of each may be quite different. Most of these operation will generate some amount of explosive or otherwise hazardous waste. It is essential to have very explicit safe operating procedures (SOPs) for each operation, to generate and adhere to good work practices, and to maintain both the facilities and equipment in a clean and fully functional condition.

Any temperature chamber used to condition explosive thermally must have limit switches that will shut down the chamber if the temperature rises a few degrees above the set point. Usually, there are two controls: The first control sets the desired temperature cycle, and the second control sets the shutdown point.

Pressing of explosives is always done behind a shield. In production facilities, the air or hydraulic supply to the press is cut off when the shield (door) is in the open position. In some facilities removable shields are often used. The shields are placed in front of the press, and the operator works in front of the shield with his/her arms around it to set the sample into the press. To operate the press, both hands are required to press two control buttons or operate two switches under the table that holds the press.

It is impossible to prevent all accidents; therefore, it is prudent to minimize the amount of explosive used in each operation and to limit the number of personnel exposed.

8.2.1 Explosive and Personnel Limits

The quantity of explosive at an operation location should be the minimum necessary to carry out a particular operation in a safe and efficient manner. When practical, this quantity should be subdivided and separated to prevent propagation of detonation. In no case should the quantity of explosive permitted in an operating building exceed the maximum permitted by the quantity-distance criteria.

All rooms, bays, and buildings containing explosive should have placards posted in a conspicuous location showing the maximum allowable amount of explosive. The allowable explosive limits will differ from one facility to another, and should be determined by a careful and thorough analysis of the particular operation. The number of personnel at an operating location should be the minimum consistent with safe and efficient operation. In establishing personnel limits, the following principles should apply:

1. Jobs not necessary to the performance of a hazardous explosive operation should not be performed in the same location as the hazardous operation. Personnel not needed for the hazardous operation should not be allowed in hazardous locations.
2. Personnel limits should allow for necessary supervision and transients.
3. Sufficient personnel should be available to perform a hazardous operation safely and, in the event of an accident, to be able to obtain help and aid the injured.
4. No person should work alone performing explosive activities that have a high risk of injury. Prompt and easy communication with other employees should be provided. The safe operating procedure (SOP) and the facility manager should specify which explosive activities may be performed alone.
5. All rooms, bays, and buildings containing explosives should have placards posted in a conspicuous location stating the number of operating and non-operating personnel allowed.

8.3 Good Work Practices

When working with explosives or any hazardous material, adequate work space should be required. Clutter in the work area or the walkways should be eliminated. Work benches should be cleared and the surface wiped before assembling an article or an experiment; this is not only for safety, but also to protect the article or experiment from contamination.

In areas where solid bare explosives are handled, it is desirable for the floors to be cushioned. This is not always possible; so extreme care should be taken when handling solid bare explosive. Nonmetallic conductive floors provide some protection against shock impact.

Work areas are not to be used for storage. Only those explosives required for preparation of an explosive article or experiment or a test should be in a work area. Before any explosive work is done, all personnel should read and thoroughly understand the safe operating procedure. If a particular operation is not covered by the SOP, an addendum should be written to provide the appropriate procedures.

All work areas should have adequate space to provide for a safe and efficient operation. Working with explosives should be carefully planned before starting a project. For example, before starting an assembly where mild detonation fuse (MDF) has to be cut in length, wash down the work space, repeat a washdown after the operation, and dispose of the cleaning residue in the proper containers. Adherence to established explosive and personnel limits for the work area is required. No personnel should work alone while performing explosive activities. The person performing the operation should always be monitored by a nonparticipating person. This can be achieved by a person observing from safe predetermined distance or via a TV monitor.

Personnel should not eat, drink, or smoke while working with explosives. Work breaks may be taken in a designated area, usually specified by the facility SOP.

8.4 Maintenance

Explosive work areas, laboratories, storage, and any structure that may contain explosives should be kept clean and orderly. To maintain a safe, clean area, a regular cleaning schedule should be followed. Never use combustible solvents when wiping up spills; do not use sweeping compounds containing wax to clean conductive floors. If a large explosive spill should occur in a work area, consult the appropriate material data sheets and clean up only according to the prescribed procedure.

When a piece of equipment breaks or wears out, it should be repaired or replaced immediately. Major repairs on equipment should not be performed in the explosive area if possible. If not, the explosive should be removed from the immediate area. Maintenance should only be performed by qualified personnel, and maintenance records should be kept. When the maintenance or repairs have been completed, the equipment should be independently inspected for proper operation and the area for cleanliness.

When a piece of equipment used in explosive work requires disassembly for repair or maintenance, it should be cleaned of any collected residue. All the screw threads and joints, where friction may occur, should be soaked in a compatible liquid lubrication before disassembly.

If there is a requirement to have a heat-producing device in an explosive area, a special SOP and permit should be required. The permit should state the location, time, duration, purpose, and any details pertaining to safety and fire-fighting equipment. The permit should be sent through the same sign-off channels as the required SOP. Also, the permit should be limited to the time required for one particular operation and canceled immediately upon completion of that operation.

A special permit should be required when welding maintenance is being performed in or immediately outside an explosive facility. The same procedures should be followed in preparing the permit, as in the case previously described.

8.5 Explosive Waste

Explosive production operations are required to have collection systems to filter explosive dust from the air or to collect waste from machining operations. The collector bag or filter should be changed periodically and treated as explosive waste, packed in special packing crates, and sent to an area where it can be destroyed properly. Collection systems for explosive machining waste should consist of a water supply that washes and cools the machined, explosive chips. The water washes the machined, explosives chips to a collection tray on the machine, through a discharge tube into a trough, to an outside filter collection bag. Then the contaminated water is allowed to collect in a sump so that the fine explosive particles may settle out. This system should be cleaned every 30 days or at specified intervals, and the collection bag packaged and sent for destruction.

All explosive laboratories should have an explosive waste procedure for scrap or damaged or reject explosive articles. Explosive residue from cuttings or assemblies should be wrapped in aluminum foil and marked to show the type of explosive and the approximate weight. This should be placed in a proper storage container and disposed of at regular intervals. Explosive articles are treated in the same manner, except that they are placed into a separate container and marked accordingly.

8.6 Spills and General Cleaning

A solvent that can dissolve explosives should not be used for cleaning or decontamination. The solvent can flow into small cracks or crevices and create a problem, especially if the solvent is not compatible with the explosive. When the solvent evaporates, the explosive residue is left coating the inaccessible depths of the crack. To clean small spills or small amounts of explosives, use a solution of 25% alcohol and 75% water. The cleaning residue from this operation should be placed in nonsparking containers and properly labeled. Periodically the waste material containers should be sent to a waste disposal area.

When cleaning a large area or decontaminating a piece of equipment, water or a water-steam mixture should be used. Solvents that have been tested for compatibility with an explosive may be used if necessary and when the proper procedures are followed. Only clean cloth rags, paper wipes, and approved nonmetallic brushes or scrapers should be used to remove explosives from equipment, work surfaces, or floors.

Each individual has the responsibility to clean any explosive residue that was distributed due to assembling or testing of their experiment or explosive article. However, if a large explosive spill should require a major cleanup of an area, the facility manager and safety engineering officer should be consulted with an SOP written to cover the cleanup operation. While working with explosive articles where small amounts of powder may fall out, the powder should be washed from the workbench using paper towels and a mixture of 25% alcohol and 75% water. Never use a solvent that will dissolve the explosive residue.

8.7 Explosive Handling

Flame- or heat-producing devices are not allowed in or around any explosive facility, unless a special permit is obtained through the facility management. This also applies to an open flame outside and for some distance away from facilities where maintenance crews may be required to use weed burners to control vegetation.

An operator has to consider all possible hazards when handling or assembling an explosive article, test item, or experiment. This includes the type of explosive, the material that will come in contact with the explosive, how the explosive is held into the assembly, and the location in the facility where the operation is being conducted. There are a number of questions that should be asked. For example: Does the assembly area have the proper equipment for the type of explosive that is being handled? Is the explosive material or article friction sensitive? If it is, how sensitive? Will the holding fixture apply excessive pressure on the explosive? Is the material that is being used to hold the explosive in place compatible with the explosive, etc.? An SOP is the first step toward identifying all these possible hazards. An SOP requires evaluation of the explosive operation and allows the personnel who will be conducting it an opportunity to critique the operation.

If possible, always work in a static-free area while handling explosives. Each explosive facility should have an atmospheric electrostatic gradient monitoring system, which should be used before handling explosives out of doors. Explosives should not be handled outside a building when the static gradient is above 2000 volts per meter (whether the voltage is positive or negative). This voltage is the potential electrostatic charge that is developed due to electrically charged clouds, electric storms, blowing dust or snow. Electrostatic charge can also be generated when a person walks across a carpet, pours a liquid from one container to another, or drags a fixture across a workbench. When working with electrostatic-sensitive materials or articles, every item in the explosive facility must be kept at the same ground potential, and each person should be electrically connected to that ground with a wrist strap.

8.7.1 Heavy Loads

There are weight restrictions for the manual handling of explosive packages or bare explosive. One person may lift or carry 25 kilograms of explosives if the explosives can be securely gripped. Two persons may lift or carry up to 50 kilograms if the proper lifting or handling tools are provided. When a person is handling explosives packaged in wooden crates, gloves and aprons are required to protect against splinters and sharp corners. Steel-toed safety shoes should be worn whenever a person is handling heavy loads. Personal objects such as combs, knives, and similar items that may produce a spark should not be taken into explosive areas.

When the 50-kilogram limit is exceeded, a cart or hand truck can be used. Carts used to move bare explosives should have a padded surface and tie downs with a lip or siding. The cart or hand truck should be equipped with brakes or chocking to prevent an unattended cart from moving. A hand cart should have a low center of gravity to prevent tipping.

Both metal chains and metal straps may be used as tie downs when explosives are sealed in shipping containers. They are not to be used for bulk explosives or containers that are not sealed. Tie-down bolts, rings, and straps used for explosive restraints must be adequate for the load and should be visually inspected before each use. When a strap becomes frayed, it can lose a large percentage of its strength; therefore, it should be disposed of immediately. Straps should be inspected every 6 months, and chains should be tested every 4 years. All tie-down or lifting devices should be tested to 150% of their recommended capacity.

All items used to lift explosive or assemblies containing explosives should be initially proof tested and periodically inspected to ensure that they will meet their specific limits.

The use of mechanical hoists is restricted in some areas for lifting explosives. Usually air-powered or vacuum lifts are used in explosive assembly areas. This type of equipment has to be inspected often because it can develop irregular motion with a small amount of wear or a small change in pressure.

Gasoline-powered material handling equipment may only be used in areas where the explosives are properly packaged. They should be equipped with backfire deflectors, an oil-bath air filter, and tight-fitting vented gas caps, and provision should be made to prevent fuel line damage due to vehicle vibration.

All battery-powered material handling equipment should meet OSHA standards. These standards cover type of batteries, electric cable mounts, electric cable runs, electric cable connection, and protection of cables and batteries.

All gasoline, liquid petroleum gas, and battery-powered equipment should have periodic inspections, and maintenance should be performed at specific intervals set by the responsible organization.

Whenever gasoline-driven lifting devices are used to move or lift explosives, spark arrestors, backfire-deflector, and sealed gas systems are required.

8.8 Testing and Firing of Explosives

As emphasized throughout this text, extensive planning is required before firing an explosive shot or conducting an explosive test or experiment. All details must be addressed in an SOP written explicitly for the firing operation.

Several types of instrumentation may be used to observe explosive experiments: high-speed cameras, video cameras, electronic recording equipment, and X-ray equipment. Monitoring an explosive experiment provides valuable test data and, in the event of an accident, can provide important information as to its cause. Passive explosive tests (those not requiring active instrumentation) are often observed through a viewing port or a shield. Precautions should be taken to limit the number of observers to protect against injury in the event of an accident.

Typically, explosives firing areas vary from remote sites to sites located near other nonexplosive buildings with many access routes, to very small test and firing chambers in controlled laboratory environments. In remote areas, road blocks should be set up and a guard posted to maintain controlled access. In areas where nonexplosive buildings are in close proximity or in controlled laboratories, the facility manager should maintain control. These areas should have a limited number of operators and casuals.

Most explosive experiment and test firing pads are limited to small amounts of explosives (a kilogram or less) where overpressure, fire brands, and fragments are not a concern. However, some of the remote facilities may be firing hundreds of kilograms of high explosives per shot. The fire brands should be controlled by housekeeping and maintenance, and fire trucks should be required to be on site while conducting a large explosive shot. High-velocity fragments in open areas are of great concern and must be taken into account when establishing the range limits or safe boundaries.

8.8.1 Safe Operating Procedures for Firing Explosives

Firing range SOPs should be established at each explosive facility to ensure that no personnel or transient will be exposed to overpressure or fragments and that proper safe boundaries are defined. The following guidelines should be considered and written into the SOP:

1. Control personnel access to all gates, doors, or roads.
2. Establish a secure firing procedure.
3. Use interlocks or locks with special keys.
4. Account for all personnel and visitors during testing.
5. Check and clear area before testing (telephone repairmen, road maintenance crews, etc.).
6. Coordinate with all local test facilities.
7. Consider any special requirements for the facilities (such as watching for aircraft when firing large explosive tests).
8. Determine the electrostatic potential gradient.
9. Establish detailed misfire procedures.

When setting up a test matrix, as much work as possible should be done before the explosives arrive at the test facility. All safety devices should be checked and tested before the explosives are installed. The instrumentation lines should be tested before the explosive installation to ensure there are no electric current paths to the fixtures holding the explosives. After the explosives are installed, only those persons that require access should be allowed at the ground zero station. Before the test, any electrical instrumentation circuits that are in proximity to the explosive charge should be checked remotely. The firing circuits should be tested from a bunker or safe firing location just prior to the test with no personnel at ground zero.

The electrostatic potential gradient reading should be taken at regular intervals during the test setup, and the range safety officer should make visual observations of the surrounding skies for approaching electrical storms. Explosives assemblies may be left unattended at a facility only when their security is assured and warning signs are displayed at all entrances to the area.

When the explosive charge is ready to be fired, the range safety officer should notify the road guards, announce the pending shot on the radio, and give a warning signal (three short siren blasts is customary). After the shot is fired, the range safety officer should give one long siren blast, wait at least several minutes, then enter the area for a visual inspection before allowing anyone else to enter the area.

If there is a misfire, the range safety officer should follow the misfire procedures written in the SOP. Usually, the procedure is to check the fire lines and, if the problem can be corrected at this point, try a second time to fire the shot. If the second try fails, the firing sequence is followed in reverse up to the point of removing the detonator. No one should physically check or approach the shot explosive assembly for at least 30 minutes and in some cases for several hours or overnight, depending on the particular situation.

8.8.2 Range Limits and Safe Boundaries

When planning an explosive shot, it is imperative to know or anticipate the effects that will be produced. Personnel should not be intentionally exposed to overpressure exceeding 0.25 psi. Excessive pressure upon the human body can cause severe injury or death. Figure 8.2 shows the probability of fatal injury (from Bowen et al.; see bibliography) along with the probability of hearing injury due to the pressure and time duration of a single blast pulse.

The range required to ensure safety from blast pressure can be estimated from the blast scaling chart previously seen in Section 5.3. When using this chart, it should be remembered that due to reflection from the surface in a ground burst, the pressures derived are equivalent to those that would be obtained from twice that amount of explosive if the shot were fired in a free field (the chart in Section 5.3 is for a free field).

Example 8.1

We wish to know the distance at which the peak overpressure from a charge of TNT would drop to 0.25 psi from a surface burst as a function of the weight of explosive used. From Figure 5.7 we find that the scaled distance that yields 0.25 psi is approximately 95 $ft/lb^{1/3}$. We know that since this is a surface burst, the yield is equivalent to that from twice the amount of explosive used, so:

$$Z = R/(2W)^{1/3} = 95$$

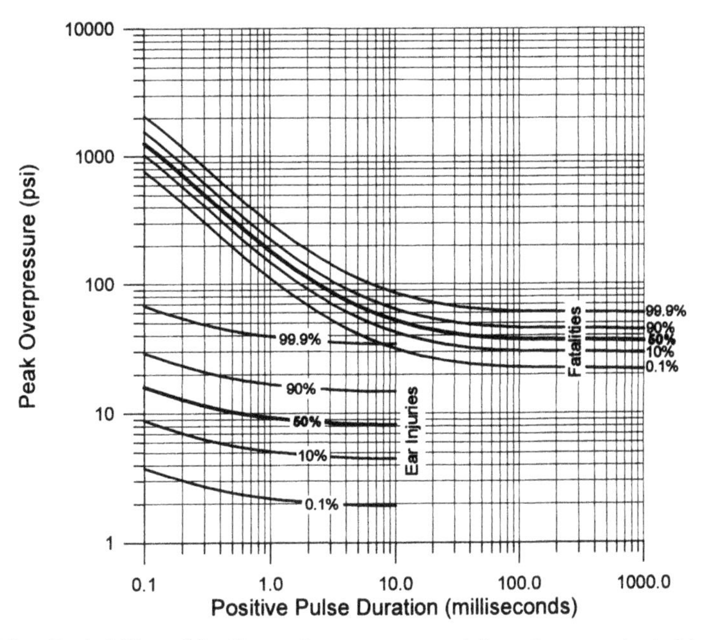

Figure 8.2. Probability of fatality and permanent ear injury due to a single blast wave as a function of peak pressure and pulse duration.

and rearranging this gives

$$R = 95(2W)^{1/3} \quad \text{or} \quad R = 120W^{1/3}$$

where R is the distance from the charge in feet and W is the weight of the charge in pounds.

The threat from fragments generated by an explosion most often extends to much greater distances than that from blast pressure. Primary fragments, those formed from a metal casing in contact with the explosive, can have velocities of several thousand feet per second. For fragments of the same shape, those having higher mass will travel further. The angle of launch also determines the maximum range of flight of a fragment, as does the fragment shape. When estimating the maximum safe range to protect from fragments, one should always be as conservative as possible. An analysis done by R. H. Bishop in the late 1950s (*Maximum Missile Ranges from Cased Explosive Charges*, SC-4205(TR), Sandia National Laboratories, July 1958) made a number of conservative assumptions about the formation and flight of fragments and produced a chart for estimating the maximum ranges. Part of that chart is reproduced here as Figure 8.3, where the maximum range for cube-shaped tumbling fragments of steel and aluminum is shown. Assuming that the maximum size fragment estimated for a particular

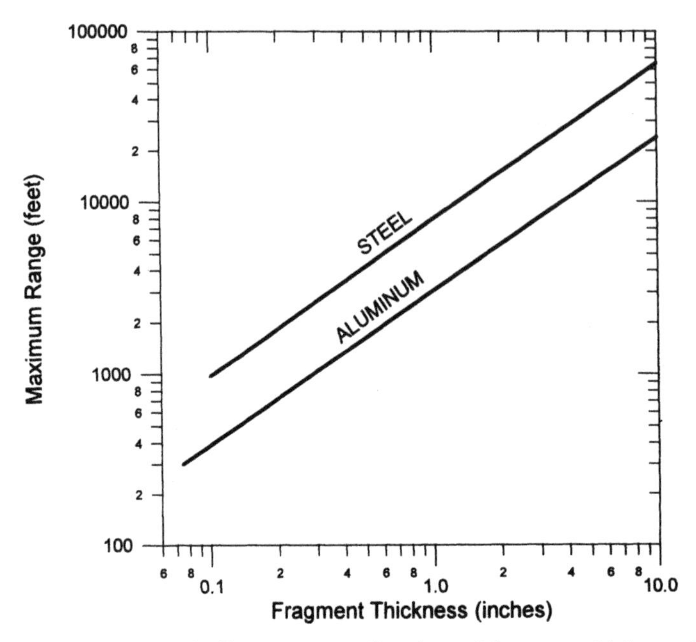

Figure 8.3. Maximum missile range as a function of fragment thickness for cubic-shaped tumbling fragments (after Bishop).

shot, regardless of its shape, is a cube of equivalent weight is a conservative way to estimate maximum fragment range.

In many test ranges as well as with blasting operations, mobile radios are used for local communications. In blasting and remote field experiments there are sometimes occasions where the operations may be near other radio antennas such as AM or FM broadcasting relays and TV and UHF relay transmitters. The transmitted signals can be picked up by firing leads and lines (which can act as receiving antennas), and if the signal is strong enough, the current generated may fire the initiator at the end of that line. Firing lines when open act as dipole antennas. Shorting the lines does not really help in this instance because a shorted line will act as a loop antenna.

Minimum safe distances from different types of transmitters have been established and are given in many explosives safety manuals, including those listed at the end of the previous chapter. Tables 8.4, 8.5, and 8.6 are charts for estimating safe explosives working distances from various kinds of transmitters.

8.8.3 Safety Considerations for Arming and Firing Systems

Every explosive firing system is referred to as an Arming and Firing system (A&F) regardless of the size or type of fireset. Before an explosive device can be fired, it must first be armed (prepared for accepting a fire signal). The arming

Table 8.2 Minimum Safe Distances between Mobile RF Transmitters and Electrical Blasting Operations

Transmitter Power (watts)	Minimum of Safe Distances (feet)				
	MF[a]	HF[a]	VHF(1)[a]	VHF(2)[a]	UHF[a]
5[b]	—	—	—	—	—
10	40	100	40	15	10
50	90	220	90	35	20
100	125	310	130	50	30
180[c]	—	—	—	65	40
250	200	490	205	75	45
500[d]	—	—	209	—	—
600[e]	300	760	315	115	70
1,000[f]	400	980	410	150	90
10,000[g]	1,200	—	1,300	—	—

[a]

MF	1.6 to 3.4 MHz	Industrial
HF	28 to 29.7 MHz	Amateur
VHF(1)	35 to 44 MHz	Public use
	50 to 54 MHz	Amateur
VHF(2)	144 to 148 MHz	Amateur
	150.8 to 161.6 MHz	Public use
UHF	450 to 460 MHz	Public use

[b] Citizens-band radio (walkie-talkie), 26.96 to 27.23 MHz and cellular telephones, 3 watts power, 825 to 845 MHz; minimum safe distance, 5 feet.

[c] Maximum power for 2-way mobile units in VHF, 150.8 to 161.6 MHz range, and for 2-way mobile and fixed-station units in UHF, 450 to 460 MHz range.

[d] Maximum power for major VHF 2-way mobile and fixed-station units in 35 to 44 MHz range.

[e] Maximum power for 2-way fixed station units in VHF, 150.8 to 161.6 MHz range.

[f] Maximum power for amateur ratio mobile units.

[g] Maximum power for some base stations in 42 to 44 MHz band and 1.6 to 1.8 MHz band.

Table 8.3 Minimum Safe Distances Between RF Transmitters and Electrical Blasting Operations

Transmitter Power (watts)	Minimum Safe Distances (feet)	
	Commerical AM Broadcast Transmitters	HF Transmitters Other than AM Broadcast
100	750	750
500	"	1,700
1,000	"	2,400
4,000	"	4,800
5,000	850	5,500
10,000	1,300	7,600
25,000	2,000	12,000
50,000[a]	2,800	17,000
100,000	3,900	24,000
500,000[b]	8,800	55,000

[a] Present maximum power of U.S. broadcast transmitters in commercial AM broadcast frequency range (0.535 to 1.605 MHz).

[b] Present maximum for international broadcast.

Table 8.4 Minimum Safe Distances Between TV and FM Broadcasting Transmitters and Electrical Blasting Operations

Effective Radiative Power (watts)	Minimum Safe Distances (feet)		
	Channels 2–6 & FM	Channels 7-13	UHF
up to 1,000	1,000	750	600
10,000	1,800	1,300	600
100,000[a]	3,200	2,300	1,100
316,000[b]	4,300	3,000	1,450
1,000,000	5,800	4,000	2,000
5,000,000[c]	9,000	6,200	3,500
10,000,000	10,200	7,400	6,000

[a] Present maximum power, Channels 2–6 and FM.
[b] Present maximum power, Channels 7–13.
[c] Present maximum power, Channels 14–83.

procedure may consist of a relatively simple action such as connecting the initiator to the fire cable; however, the process is often more complex. All firing systems should have an arming and firing sequence, which is performed at the firing console after the initiator is connected to the system. This system utilizes an arming-control key. The basic, normal procedure should be as follows:

1. Remove power from the firing set in the console.
2. Remove the A&F key; the person who will fire the shot should secure this key.
3. Assemble the shot or experiment.
4. Clear the firing pad or area.
5. Mount the initiator to the experiment and connected it to a shorted firing cable.
6. Connect the cable to the firing system's circuit.
7. Give warning to all personnel involved in the test and to the general area.
8. Arm and fire the experiment.

Whenever firing an explosive shot, the fireset should be connected to an interlocking system that is connected to the critical access area. Interlocks are

Table 8.5 Certified Meters Used by the U.S. DOE National Laboratories

Name/Manufacturer	Model No.
Alinco Igniter Circuit Tester	101-5BFG
Fluke	8012 A/AD(SE3065)
Digital Voltmeter (Sandia Nat'l Labs)	PT4030
Valhalla Igniter Circuit Tester	4314A

designed to shut down the firing system if, for example, a door is opened that might give any personnel access to the firing area. Some firing pads have ABORT buttons at or near the firing pad that are part of the interlock system.

In all explosive operations there should be a two-person rule. This means that there must always be a person observing all explosive operations from a safe distance. If three persons are working on an explosive assembly, a fourth person should be observing from a safe distance. The person who is the observer may be in another room or building and monitoring the operation using a sound system or a video system or both. Obviously, the reason for this rule is so the observer can report any incident that may occur and render first aid if necessary.

All firing systems should use a fail-safe concept. When the experiment is ready to be tested, a key is inserted into the safe-arm switch; the fireset relay is reset; the fireset energy source is charged or activated; and the system is now ready to be triggered. This sequence should be followed in order to send energy to the initiator. If the fireset is triggered and the initiator fails to fire, the system should have the capability to shut down automatically and not be enabled until the fireset has been manually reset.

If an explosive component fails during an experiment or test and a post mortem is required to determine the failure mode, it should be done in a remote area. There should be special areas where this type of operation can be performed.

8.8.3.1 Ignitor and Firing Circuit Test Meters

The meters used to test or check initiators and firing circuit lines and elements must be of a special kind. Standard or common ohm-meters produced relatively high currents through the test leads, as much as a half-amp for some meters on the low-resistance scales. This is sufficient to fire many initiators and types of blasting caps. Therefore special meters with restricted output current must be used.

The output current of test meters (used for firing systems and initiators) through a resistive load should not exceed 1% of the minimum all-fire current of the most sensitive initiators to be tested. Also, the current should not exceed 10% of the no-fire current rating for the most sensitive initiator to be tested. In all cases, meters used for test circuits with initiating devices shall have internal current limiters. Most such meters are limited to no more than 5 milliamps maximum output current. Any new instrument must be certified by an approved measurement standard laboratory. Shown as examples in Table 8.5 are four of the certified meters used by U.S. DOE laboratories.

8.9 Licenses, Permits, and Penalties

In the United States as well as in most other countries, it is required to be licensed or to obtain special permits for working with or dealing in explosives. The laws of each country are different but are written for the same reason, to ensure the safety and security of the public and commerce. The United States requirements are given here as an example.

Licenses and permits are required by the federal government for the use, import, manufacture, distribution, sale, and storage of explosives. In order to obtain such licenses or permits, the applicant must meet a list of requirements given in 27 CFR 55 and the Organized Crime Act, Title XI (Regulation of Explosives), also known as Public Law 91-452. Among the requirements are that the applicant: may not distribute explosives to unlicensed parties, must not have willingly violated any of the provisions of the explosives laws, must have licensed and approved explosive storage facilities, must demonstrate knowledge and familiarity with the explosives laws, be over 21 years of age, must not have been convicted of a crime punishable by imprisonment exceeding a year, must not be a fugitive from justice, must not be a user of illegal drugs or controlled substances. This is only a part of the list but demonstrates the spirit of the requirements.

Violation of the civil portions of the federal regulations of explosives can be punishable by fines up to $10,000 and imprisonment of up to 10 years for each offense depending on the particular offense. Violation of the criminal portions can be punishable by fines up to $20,000 and imprisonment for any number years if a death is involved.

In addition to federal licenses and permits, there are others required by most state and local governments, many of these also having financial liability requirements.

The use and storage of explosives also comes under the purview of a totally different set of laws and regulations, these dealing with environmental protection, occupational health and safety, noise abatement, and public safety to name only a few.

This section certainly does not give all the myriad safe, responsible, and legal requirements for working with explosives; it is mentioned only to give the reader insight into the nontechnical aspects of this exciting and challenging business.

8.10 Related Reading

1. *A Manual for the Prediction of Blast and Fragment Loadings on Structures*, DE82-000536, DOE/TIC-11268, U.S. Dept. of Energy (1980).

2. R. H. Bishop, *Maximum Missile Ranges from Cased Explosive Charges*, SC-4205(TR), Sandia National Laboratories, Albuquerque, New Mexico (July 1958).

3. I. G. Bowen, E. R. Fletcher, and D. R. Richmond, *Estimate of Man's Tolerance to the Direct Effects of Air Blast*, HQ DASA, DA-49-146-xz-372, Washington, D.C. (October 1968).

4. *Safety Library Publications* from the Institute of Makers of Explosives (IME), 1120 19th Street NW, Suite 310, Washington, DC 20036.
   - 4.1 Suggested Code of Regulations (1/85).
   - 4.2 Glossary of Commercial Explosives Industry Terms (2/91).
   - 4.3 Safety Guide for the Prevention of Radio Frequency Radiation Hazards in the Use of Commercial Detonators (12/88).

Appendix

A.1 Conversion Factors

In the field of explosives, a myriad of units are used. Many sources of data report in different units. An example is pressure, which is given in psi, atmospheres, bars, kilobars, pascals, and gigapascals. The following will be of help in juggling units often used in working with explosives.

Weight

7000 grains (grn) = 1 pound (#)
453.6 grams (g) = 1 pound
16 ounces (oz) = 1 pound
1 ton (short) = 2000 pounds
1 ton (long) = 2240 pounds
14 pounds = 1 stone

Length

25.4 millimeters = 1 inch (in.)
2.54 centimeters = 1 inch
30.48 centimeters = 1 foot (ft.)
1 meter (m) = 39.37 in.
1 angstrom = 10^{-10} meters
220 yards = 1 furlong

Volume

16.39 cm^3 = 1 $in.^3$
1000 cm^3 = 1 liter
1 liter = 61.02545 $in.^3$
1 ft^3 = 28.316 liters
3.785 liter = 1 gallon (US, liq.)
4.404765 liter = 1 gallon (US, dry)
4 quarts (US) =1 gal (US)
2 pints = 1 quart
1 gal (US) = 231 $in.^3$
1 ft^3 = 7.481 gal (US)
16 fluid ounces = 1 pint
a fifth = 0.8 quarts

Pressure

1 psi = 2.036 in. Hg (0°C)
1 atmosphere = 14.7 psi
1 bar = 14.5 psi
1 atmosphere = 760 mm Hg
1 atmosphere = 29.921 in. Hg
1 bar = 0.986 atmosphere
1 psi= 6897 pascals
1 bar = 100 kilopascals
1 gigapascal = 10 kilobars
1 pascal = 1 newton/m^2

Temperature

1°Celsius (C) = 1.8°Fahrenheit
1°Celsius = 1°Kelvin (K)
1°Rankine (R) = 1° Fahrenheit (F)
0°K = absolute zero
0°R = absolute zero
T°F = 1.8 × T°C + 32
T°K = T°C + 273.16
T°R = T°F + 459.7

Energy

1 watt-second = 1 joule (j)
1 calorie (cal) = 4.184 joule
1 BTU = 252.15 cal
1 erg = 10^{-7} joule
1 joule = 0.73755 ft-#

Power

1 watt = 14.34 cal/min
1 watt = 44.25 ft-#/min
1 kilowatt = 56.87 BTU/min
1 kilowatt = 1.341 horsepower
1 horsepower = 550 ft-#/sec
1 watt = 1 j/sec

Velocity

1 mm/microsec = 1 km/s
1 km/s = 3281 ft/s
1 mile/h = 1.609 km/hr

Density

1 g/cm^3 = 0.0361 #/in.^{3}
1 g/cm^3 = 8.345 #/gal

A.2 Frequently Used Constants and Relationships

Volume of 1 g-mole of gas (0°C, 1 atm) = 22.414 liters
Volume of 1 pound mole of gas (32°F, 14.7 psi) = 359.05 cubic feet
Gas law constant, R = 1.987 (cal)/(g-mole)(K), = 0.08205 (liter)(atm)/(g-mole)(K)
Avogadro's constant = 6.024×10^{23} molecules per g-mole
Acceleration due to gravity at mean sea level = 32.2 ft/s^2

A.3 Multiplier Nomenclature

pico = $\times 10^{-12}$
nano = $\times 10^{-9}$
micro = $\times 10^{-6}$
milli = $\times 10^{-3}$
kilo = $\times 10^{3}$
mega = $\times 10^{6}$
giga = $\times 10^{9}$

Index